Searching African Skies

Searching African Skies

Sarah Wild

First published by Jacana Media (Pty) Ltd in 2012
Second impression 2013

10 Orange Street
Sunnyside
Auckland Park 2092
South Africa
+2711 628 3200
www.jacana.co.za

© Sarah Wild, 2012

All rights reserved.

ISBN 978-1-4314-0472-8

Cover design by publicide
Set in Arno Pro 12.5/14.3pt
Job No. 001995
Printed by Mega Digital (Pty) Ltd., Cape Town

See a complete list of Jacana titles at www.jacana.co.za

*To all those people who thought South Africans
would not be interested in a science book*

Contents

Preface ix
Acknowledgements xiii

Chapter 1:
 What is radio astronomy? 1

Chapter 2:
 The origins of radio astronomy in South Africa 4

Chapter 3:
 The history of astronomy in South Africa 20
 *The amaXhosa and their understanding of
 the night sky* (Temba Eric Matomela) 34

Chapter 4:
  Southern African Large Telescope 38

Chapter 5:
  Back to basics 54

Chapter 6:
  The KAT-7 and the MeerKAT 63

Chapter 7:
  African SKA.. 88

Chapter 8:
  Challenges 99
  Internet connectivity for South Africa's SKA bid
  (Toby Shapshak) 115

Chapter 9:
  The African VLBI Network................................ 120

Chapter 10:
  Benefits... 135

Chapter 11:
  SKA science... 152

Chapter 12:
  Looking to the future....................................... 174

Preface

"So what is this SKA thing anyway?" That is the most common question I'm asked at parties when someone falls back on the "What do you do?" icebreaker. When I say I'm a science and technology editor, this is inevitably met with some shoulder-surfing, the act of scanning the crowd over someone's shoulder looking for better conversation. This is because South Africa is rife with science stigma, the idea that science isn't relevant to a developing country, that science means laborious mathematical equations, and that science should stay out of everyday conversation – especially at parties.

And then the Square Kilometre Array (SKA) came along, and finally people had a question to fill that awkward oh-dear-I'm-talking-to-a-geek silence in the conversation. By now, many

South Africans have heard about the SKA and know that we were bidding against Australia to host it, and that it will be split between the two countries. Since our competitor was Australia, South Africans thought we should win on principle – we're sick of Australia denting our national pride in rugby and cricket – even though those who'd heard about it didn't actually know what it was. This is why I sit with the loaded question, "What is the SKA?"

I finally have their attention, so I don't want to delve into the origins of the universe or gravitational radiation. That will just start them shoulder-surfing again. So, I say: "It will be the biggest scientific experiment on Earth – a radio telescope that will be able to detect signals from the beginning of the universe." The person nods with satisfaction; they now have an answer that they can weave into another conversation.

Luckily, I work for a newspaper, which means I have a platform to tell people about this exciting project, but then there are the dual pressures of newspaper space and newsworthiness. That is why this book came about: to tell the story of radio astronomy and how South Africa ended up bidding to host the largest science project in the world. There is more to it than a glib, well-rehearsed *on dit* to be flung out at parties – it is a story of the people behind the project, some of whom have dedicated their lives to curiosity about the universe, others who have fought local disinterest and international condescension to make us a real competitor in the bid, and then there are those whose lives will be irreversibly changed by South Africa's radio astronomy plans. Some of them haven't even met each other, but their lives are linked by a radio telescope, and by the fact that their stories are seldom told.

When I first decided to write a book about South Africa's bid to host the SKA, people brushed me off: "South Africans aren't interested in science", "No one will want to read about that", and so on. They are the same people who will then complain about the scarcity of skills in the country, and the fact that school learners don't want to take Maths and Science as subjects. They can't see that all these problems – including the perception that no one would be interested in reading a book about South African science – are connected, and it all comes back to science stigma. How can we expect our children to be interested in science if we aren't?

Many years ago, I taught Physics to school learners before I realised that I was a better journalist than physicist. The most common problem among my students wasn't the maths; it was that they thought science was boring. This book attempts – you will have to judge whether it succeeds – to show you why science is interesting and relevant, why the SKA will change your life, and that science is for everyone, not just for PhDs.

Sarah Wild
Johannesburg, July 2012

Acknowledgements

Before I started writing this book, I thought the acknowledgements section was a cunning device to get Aunt Marge to forgive you for forgetting her birthday – and that the section was so long because authors, often being rather distracted creatures, forgot a lot of birthdays.

In reality, writing a book is a mammoth task and it is impossible to accomplish without the help of people who are willing to give up their time to talk to you, answer stupid questions and reply when you check the wording of their answer. Friendships are tested to their limits because, at best, you are only able to talk about your book and, at worst, you tend to make people read multiple drafts of the manuscript; and your loved ones are obliged to put up with you when you turn into a coffee-addled bleary-eyed, and let's be honest, irrational monster.

So many thanks are deserved in many quarters. To all the people who were prepared to give up their time when I'm sure they had better things to do: Brian Warner, Patricia Whitelock, Bernie Fanaroff, Justin Jonas, Adrian Tiplady (a special thanks to Adrian for the fact that he still takes my phone calls), Frank Curtolo, Pieter Snyman, Khotso Mokhele, Rob Adam and for all those who I've forgotten to mention.

This book is a *real* science book because of the people who not only let me commandeer hours of their time, but also checked the scientific facts for me. It is not possible to extend enough gratitude for their expertise and kindness: David Buckley, Mike Gaylard and George Nicolson; Laura Bezuidenhout from Expert Virtual Assistant Services for typing up hours of interview transcripts on what was always short notice; Temba Matomela and Toby Shapshak for their contributions – which add a special something to this book and it is richer for them; my editor, Peter Bruce, for having faith in me and giving me more chances than I deserve; Rehana Rossouw for unknowingly showing me the kind of journalist I wanted to be; and, importantly, Jack Lesage for being my friend. A big thank-you must also go to the Department of Science and Technology for seeing the book's potential, and to Tommy Makhode for taking my (many) phone calls.

When I first pitched the idea of a popular science book, people – once again – told me that South Africans weren't interested in science. Thank you to Russell Clarke, Sean Fraser and Jacana Media for being as excited by the book as I was, and for making my idea a reality.

My gratitude, too, to Cassidy Parker for all the tea and for perhaps being even more excited about the book than anyone; Simon Ferreira for restoring my faith that people would actually read a science book; Ron Fogel: I hope you're still my friend after the lengths I pushed our friendship to by making you read several drafts of the whole book; Clara Vaughan, the most mighty Ming, your friendship (not to mention the chocolate) kept me sane – well, relatively anyway. My family, and especially my mom, for listening to me talk about this book non-stop and interspersing my chattering with "you can do it".

And finally PdW. On one hand, you were my greatest support, bouncer-off of ideas and comrade-in-arms, and I couldn't have done this without you. On the other hand, damn you. You ruined my life – this was your idea!

1
What is radio astronomy?

It is 1931, and an American radio engineer sits in front of reams of paper, wondering if he is losing his mind, if there is something wrong with his instruments, or both. He adjusts his glasses and runs a hand over his prematurely balding head. On paper, his job is simple: find out what causes interference on transoceanic radio communications, specifically long-distance, short-wave communications. In real life, the task is somewhat more complicated.

Karl Jansky, an employee of the Bell Corporation, has isolated two aspects of the interference: nearby and distant thunderstorms. But there is something else. It plays through his instrumentation like a supersonic tinnitus, a static that is sometimes there, sometimes not.

He has built an antenna to measure radio waves of 20.5 MHz, rotating on four Ford Model-T tyres. That way, he can find this mysterious source of static, which seems to radiate from all directions.

If you have an image of a super-slinky, sexy, rotating apparatus, its slicked chrome-and-ceramic veneer dreamily spinning like something out of *2001: A Space Odyssey*, forget about it. This is the 1930s. Commonly called "Jansky's merry-go-round", it is a leviathan, even in its own time. Thirty-five-and-a-half metres in length, the radio-wave receiver is a series of enormous cross bars and square metal frames, and takes 20 minutes to complete a 360-degree rotation.

Close to the device, Jansky sits in a separate shed, staring at the data from his analogue pen-and-paper machine. It's there – a variable hiss that peaks once a day.

Five hundred years ago, people thought that the universe – in which the Earth was the prestigious centrepiece – comprised enormous spheres, moving together in mathematical harmony. *Musica Universalis*, the music of the spheres, was ancient philosophy, rather than actual music.

While most scholars dealt with this on a metaphysical level, Johannes Kepler – whose laws of planetary motion describe the elliptical orbits of planets around the Sun – believed that they really made music, singing like half-filled wine glasses played with a moistened finger. A mixture of religion and science, these spheres sung the music of the heavens.

According to Kepler, celestial bodies orbited each other at a fixed distance, which corresponded to a specific pitch, in the same way that a string of a certain length on a violin elicits a certain note. All these notes were weaved into a pattern of proportion, a divine pattern that determined the universe.

Now, while Kepler was on the mark about the maths, he was a bit confused about the music.

Tangentially, while the ancients were wrong about a number of things, such as a Geocentric Earth and the health benefits of leeches, they were half right about the music of the spheres.

Initially, Jansky thought the static was originating from the Sun because of the undulating nature of the maximum intensity. But it didn't fit with the 24-hour solar day – with pure obtuseness, the signal repeated every 23 hours and 56 minutes, in sync with the sidereal day, which is measured via the Earth's rotation with respect to a fixed extrasolar object.

What Jansky had in fact stumbled upon was radio waves emitted by the Milky Way.

Now the 1930s were a tough time to find funding for large experimental science projects. The Great Depression had struck the world in the late 1920s and no one had money to spend, so – although Jansky published his findings in 1933 – the discovery lay dormant for a number of years.

He tried to convince Bell Laboratories to build an even bigger antenna, but experimental physics paled in comparison to the demands of a bottom line, and Jansky was redeployed within the company.

It took decades for scientists to discover that other celestial bodies emitted radio waves at other frequencies, but the invention of Jansky's merry-go-round formed the foundation of the new discipline we have come to know as radio astronomy. Stars, galaxies, pulsars, quasars – they all emit radio waves of different frequencies.

Traditional astronomy used, and still uses, optics to see into outer space, but the range of the visible spectrum – the seven colours that the human eye can see – forms only a tiny part of the electromagnetic spectrum. Radio waves occupy a large portion of the spectrum, which means that you can detect a greater variety of wave and can gain fresh insight into the universe. By collecting and collating the data from the different electromagnetic-radiation frequency bands, we are able to map out the dark recesses beyond the Earth, like blind cartographers listening to the sounds of the universe.

2
The origins of radio astronomy in South Africa

It's all America's fault. If Australia is looking for someone to blame for its Square Kilometre Array competition, it should blame the United States. The existence of radio astronomy as a discipline in South Africa is actually thanks to the United States' National Aeronautics and Space Administration (Nasa). Until the KAT-7 was built in the Karoo in 2010, the country's first and only radio astronomy observatory was once Nasa's Deep Space Station 51, nestled in the hills of Hartbeeshoek, Gauteng. It sounds like something out of a Cold War spy movie, or at least a science-fiction thriller involving aliens and Armageddon.

Well, it was the Cold War, and the Russians had fired the first Space Race salvo in 1957 by launching the satellite *Sputnik 1* into space. In 1960, the then Union of South Africa's Council for Scientific and Industrial Research (CSIR) entered into an agreement with fledgling Nasa to host one of its Deep Space Stations in the country, to track American automated lunar and planetary probes.

Satellite and rocket launching is a risky business, and the spacecraft needs to be tracked throughout its launch. But if it's launched in the United States, the Americans don't have eyes on the other side of the world to watch its movements. So Nasa's Jet Propulsion Laboratory established three Deep Space Stations, one in South Africa, and another two in California and Australia. The stations were approximately 120 degrees apart in longitude and effectively divided the Earth's rotation into thirds, providing continuous and overlapping coverage.

Not to sound esoteric, but planetary motion also played a role in the establishment of tracking stations in the Southern Hemisphere. In the 1960s, the inner planets out to Mars all had southerly declinations, which meant that if a probe was sent to Mars, it would remain very far south for months at a time. As a result, it was only possible to continuously track spacecraft – and sometimes Mars – by having a station in South Africa rather than Europe. The Southern Hemisphere doesn't permanently have this natural advantage over the north. Planetary declinations are cyclical. Through a confluence of circumstances – lucky circumstances for South Africa – the planets were favouring Southern Hemisphere observation during the 1960s and early 1970s.

Without Nasa's involvement, it is unlikely South Africa would have had a radio astronomy programme at all. At the very least, the country wouldn't be as advanced in this field as it is today.

The people holding the purse strings in the late 1950s thought that Australia was too far advanced in radio astronomy for South Africa to be truly competitive. It is also worth noting that, just as it is today, radio astronomy is a rather expensive field to be involved in.

Australia's competitiveness dates back to the post-Second

World War period. Both South Africa and Australia were building radar systems because, as Commonwealth countries, they had developed radar in conjunction with radar specialists in the United Kingdom to defend their respective coastlines against attack. Radar, short for radio detection and ranging, works by bouncing radio waves off objects to determine their location, direction and speed, and you need both a radio transmitter and a receiver.

Radio astronomy uses similar techniques, except that it doesn't need a transmitter. Celestial objects, which generate a wide variety of natural radio signals, are the transmitters, and the receivers need to operate across a range of frequencies.

After the Second World War, there was a surplus of radar equipment and a surprising discovery had thrust Karl Jansky's cosmic radio waves into the spotlight. British physicist James Stephen Hey had been tasked with devising radar anti-jamming methods during the war. In February 1942, he received reports of severe noise interfering with the Allies' anti-aircraft radars. At first, he thought it was caused by crafty radar-jamming techniques or interference by German forces, but after some discussion with the Royal Society of London for Improving Natural Knowledge, he realised it was radiation from the Sun. But because it was wartime and "keeping mum" was the order of the day, Hey was not allowed to publish his discovery until after the war.

In this post-war period, Australia got an impressive head start in radio astronomy. The country had a number of radar specialists, and unlike countries such as the United Kingdom and the Netherlands, there was little organised academic research. While this might sound like something that might hold back academic inquiry, it isn't. It meant that people with radar skills who were interested in radio astronomy could just start tackling research problems, without it having to be coordinated into university groups and limping through an obstacle course of academic bureaucracy.

If you have a discussion with anyone in South Africa's radio astronomy community about the history of the discipline, they will tell you to speak to George. Or else they will start telling you stories, and interspersing them with comments such as, "You need to check with George about the dates. He'll remember" or "George has a good memory for how things went forward".

So when you finally meet Dr George Nicolson, he is a bit of a surprise. The grandfather of South African astronomy – who was the first employee at Nasa's deep space tracking station and the Hartebeesthoek Radio Astronomy Observatory's (HartRAO) first director – has a shock of white hair, blue eyes that twinkle as brightly as the stars he's spent his life looking at, and a mischievous smile.

The reason people will direct you to Dr Nicolson is because he was there, he lived the early days of radio astronomy. So, when asked about Australia's prowess at that time, he is unequivocal: "They were undoubtedly world leaders, in terms of publications, number of people they had working in radio astronomy and the quality of work that they were doing."

Because of this, the first president of the CSIR, Dr Basil Schonland – who also led the development of South Africa's radar system – believed the country had no chance of catching up with Australia, and should instead concentrate on something that was unique to South Africa, such as using radar to study lightning because the country has such a high incidence of the phenomenon.

Nasa's Deep Space Station 51, however, removed the impediments of funding and Australia's competitiveness. The constrained budget was no longer an issue because Nasa was bankrolling the infrastructure. And, in the Space Race, US pride was on the line, so money didn't seem too difficult to come by. The space agency built a 26-metre dish in Hartbeeshoek's verdant hills, far away from possible interference. It operated in the 960 MHz frequency range, designed for signals with a wavelength of about 30 centimetres. The only way to get there, from either Pretoria or Johannesburg, was along dirt roads.

Concurrently, Rhodes University in Grahamstown had started its own mini radio astronomy programme, by using conventional shortwave communications receivers and low-cost high-frequency wire antennas, which further diluted the cost threat of the country becoming involved in radio astronomy.

Nasa provided the skeleton for radio astronomy in South Africa, and it was literally a skeleton. What now appears to be a handsome parabolic dish comprised, in those days, a seemingly flimsy wire mesh surface. "When you looked at the dish, you just looked straight through it," Dr Nicolson says. "It was quite a spidery structure and it went through various upgrades over the years."

But, in an example of why you shouldn't judge a book by its cover, that "spidery" telescope also housed state-of-the-art Nasa technology. The deal brokered between Nasa and the CSIR – the council's National Institute for Telecommunications Research in Johannesburg was responsible for the tracking station – stated that as long as it didn't disrupt the dish's primary tracking function and duties, the dish could be used for radio astronomy.

Not that there were many people who could use it for radio astronomy – there was only Dr Nicolson. He had originally worked for the CSIR, but was transferred to the Hartebeesthoek facility as "basically their first employee".

"The one thing the director of our institute [Dr Frank Hewitt] impressed on me right from the beginning was that it was going to be a one-man programme. He said that I had to be careful to choose projects that people with greater resources couldn't complete in a shorter period of time," Dr Nicolson says.

So, as part of his MSc degree, he began a survey of the Southern Galactic plane at 960 MHz. "So I initially started off by completing the survey of the southern Milky Way galaxy, preparing for it in 1961, building the equipment in 1962 and carrying out and analysing all the observations in 1963." A more extensive survey had been undertaken by the Owens Valley Radio Observatory in California, but because of its geographic location, observations from the Southern Hemisphere were needed to fill in gaps.

Back then, in 1961, radio astronomy was worlds away from what it is now. These days it is such a complex discipline that niche specialities have developed: data management, computer interfacing, data analysis, telescope construction, receiver technologies.

Dr Nicolson takes a moment to think, before commenting on the differences between nascent and modern radio astronomy. "In the earlier days of radio astronomy, people had to have a range of skills. They were mostly physicists or engineers rather than astronomers, so they had to learn astronomy along the way. As radio astronomy developed, people became less dependent on having technical skills to do radio astronomy. Today, most of the world's radio astronomers know very little about the technical details. They rely on the engineers and those astronomers who have technical knowledge and who have developed systems to provide them with a working instrument. They just get the data that streams out."

These days, radio astronomy comprises a large number of specialists, who work in niches to synthesise this complex beast. Back in the 1960s, radio astronomers built their own instruments, and if something didn't work they had to fix it.

Aside from the fact that Nasa provided the 26-metre dish, there were other definite benefits to being affiliated with an organisation that was trying to one-up its Russian competitors. For one thing, the relationship brought with it the most cutting-edge and sensitive gadgets available.

In 1964, Deep Space Station 51 changed its operating frequency from 960 MHz to 2290 MHz. While Nasa did this to improve its tracking capability, South Africa got caught in the slipstream of four-times-more sensitive equipment. The amplifier Nasa installed gave the country's radio astronomy community proportional advantage because it made the dish more sensitive in some respects than any other radio telescope in the world operating at that frequency range.

However, the receiver on the telescope also needed a facelift and a new noise-adding radiometer, which – through necessity because there wasn't really anyone else to do the job – was built by Dr Nicolson.

With a new operating frequency and machinery so advanced it should have worn a space suit and had a spacecraft powered by cold fusion, the lone radio astronomer/engineer needed a new project. So Dr Nicolson decided to observe quasars, which had recently been discovered. A quasar, otherwise known as a quasi-stellar radio source, is a region at the centre of a galaxy, usually a very young galaxy, and is one of the most luminous and energetic objects in the universe. Scientists believe that it surrounds a supermassive black hole that is so dense and has such a strong gravitational pull that not even light can escape it. Quasars can be up to 100 times brighter than an entire galaxy, brighter than all the Milky Way's stars packed into a small region.

Because of its location and sensitivity, the telescope also played an important part in the global Very Long Baseline Interferometry (VLBI) network, a radio astronomy technique that involves observing a single object through several telescopes in different parts of the world simultaneously, so that all the telescopes act as one big telescope. It filled in a longitudinal gap in the Southern Hemisphere, and was the only radio telescope on the African continent (see Chapter 9: The African VLBI Network).

☽

And so 12 years passed in relative peace, with different projects, new spacecraft to track and radio astronomy projects sneaked in when the antenna was idle. It should be noted that it is difficult for the sleepy hills of Hartbeeshoek to be anything but peaceful.

The dish nestles between the Hartbeeshoek hills, and is the only really jarring sight in a bucolic paradise. But outside the confines of South Africa's increasingly isolationist borders, trouble was brewing. We're now in the mid-1970s, and the international community had turned its weighty displeasure at the apartheid regime into sanctions, and it was becoming increasingly difficult for Nasa to justify its interaction with a state tarnished by human rights violations. By now, the planets had turned their backs on the Southern Hemisphere, and the argument for Deep Space Station 51 as vital because of South Africa's geographic location was paling in comparison to international pressure. Allegedly there were even attempts in the United States Congress to withhold Nasa's funds unless it withdrew from the pariah state.

So it was that the space agency decided to pull out of South Africa in 1973. Unfortunately the people working at the tracking station first found out about the decision in the *Rand Daily Mail* (which would later morph into the *Mail & Guardian*). There's nothing quite like discovering something personally life-altering in a newspaper, especially when it's been given the space of about eight postage stamps.

"The CSIR staff woke up having worked on the tracking shift the night before at Hartebeesthoek, turned on the news and heard that the place was closing down," Dr Nicolson says, smiling wryly as he remembers it.

As proof that deadlines are cowards and hunt in packs, the announcement also came on the day that proposals for new research funding were due. A hurried meeting between the deputy president of the CSIR and the director of the tracking institute resulted in budgets being scribbled on the back of an envelope.

When Nasa pulled out, it left redundant equipment behind, which was great for starting up a radio telescope but hardly earth shattering. Something that was shiny and new in 1964, that made your coffee while tying your shoelaces, was somewhat decrepit by 1974. But, as Dr Nicolson says, "It wasn't the latest and

11

greatest, but stuff that we could use to get a telescope working."

Virtually overnight, South Africa had its own radio astronomy observatory, even though it didn't have money to do things such as buy new equipment, and salaries were drawn from a shoestring budget. The observatory could also be used by radio astronomers at South African universities. Rhodes University had already been dabbling in radio astronomy for a number of years, building their own receivers and making inroads into what was still a very niche discipline.

A radio telescope isn't the same as a tracking dish. Spacecraft transmit information on carrier waves; the dish receives those waves and the information contained in them needs to be unpacked and analysed. The bandwidth (which in this case means the frequency range) of the carrier wave is matched to the receiver bandwidth. Radio astronomy doesn't work like that. There are a number of disciplines within the field, such as spectroscopy, which is when an astronomer looks for a specific substance, atom or molecule, for example hydrogen, which radiates at a particular frequency.[1] This is a type of radiometry: when an astronomer surveys the sky at a particular frequency and maps all the objects that they see. But objects such as the Sun, the Milky Way and quasars emit a broad spectrum of frequencies, from low frequencies all the way up to infrared, visible light, ultraviolet, X-rays and even gamma rays. As a result, you need receivers that have a relatively wide bandwidth at various frequencies. So while the main – and most expensive – infrastructure was there, almost everything else had to be built. Luckily, in those days radio astronomy required a number of skills, engineering being one of them. Nasa had left the amplifier, but the staff then had to build receivers to fit with the amplifier,

---

[1] It is the substance rather than the actual celestial object that emits radio signals. You often say that an object is emitting signals, but that's only because the star or galaxy has a high concentration of a particular atom or molecule.

which, although a decade old, was still an impressive piece of electronics.

The observatory officially began operating on 1 March 1975, with two astronomers – Dr Nicolson, who had become the director, and Dr PJ Harvey (not the famous singer) – two observing assistants and seven technical support staff. Nasa had left in its wake equipment and expertise that enabled the observatory to begin operating relatively quickly at a single frequency.

But a radio telescope that can only receive one frequency is a rather pathetic telescope, a little like having a black-and-white TV with a single channel. Proposals were thus sent off to expand its receiving capacity to include wavelengths of 18 and six centimetres, to complement its existing 13-centimetre wavelength receiver. After this got the nod, it was decided that all of the receivers should be available concurrently. If you've seen a receiver, this makes sense. The 18-centimetre receiver is about the length of a grown man's torso, jam-packed with electronics. It also has to be cooled so that the radiation from the receiver doesn't interfere with the radio telescope. You would definitely have drawn the short straw if you had to be the one who had to climb onto the dish a number of times a day (and night) to connect the necessary receiver in the freezing nose of the telescope.

Now a word on radio telescopes in the 1970s. While technology has made life easier in many respects and many arduous tasks have become automated and we don't notice that computers are doing something that would have taken us hours, we've lost a bit of the romance. Back in the early days, you literally drove a radio telescope. It wasn't like driving a car with a steering wheel, but you turned hand controls – like, if you'll excuse the pun, the knobs on a radio – to drive the telescope north, south, east or west. You could make it go faster or slower in either direction. Once you'd achieved the correct coordinates, you'd lock the telescope on that axis, and then fiddle with the other one. Once fixed, you would let the Earth's rotation do the rest of the work for you, allowing you to scan an object in the sky. These observations are called "drift scans".

While driving a telescope may sound romantic, and relying on digital computers delightfully hands-off, the intermediate stage sounds like hell. In the semi-automated system, the day's coordinates were punched onto a paper tape and fed into the control system. It was different every day because there were always different objects to observe. There was also a tape that effectively cleared the slate and returned the telescope to a known object in the sky, the first point of Aries. "It's the equivalent of 0 latitude and 0 longitude," Dr Nicolson says. Before a day's observation, that tape would be inserted into the control system, as a reference point, and then you'd embark on the day's radio source hunting, starting the telescope and letting it track the source across the sky.

While having engineers as part of the staff was useful for building new equipment, the long-standing association with Rhodes University was also a source of innovation. Rhodes became affiliated with the observatory in the early days of its existence. Before that, the university's Physics department had built – and was using – a two-metre dish, operating at 22 GHz. This provided fertile training ground for would-be radio astronomers, and when the tracking station became an observatory, it was decided that universities would get a 20% time allocation and that postgraduate students would be able to use it for their research.

Fortunately for Rhodes, there were no other universities involved in radio astronomy. The university had been using their home-built two-metre dish, so when they were told they could now use a 26-metre dish, it must have been like wishing for a toy car for your birthday and getting a monster truck with a full tank of petrol.

The present MD of HartRAO, Mike Gaylard, was a Rhodes

MSc student at that time. "I'd been working on a two-metre dish until that point," says Dr Gaylard, "so the opportunity to work on a 26-metre one was irresistible." But at this time there was only one receiver on the dish – Nasa's 13-centimetre receiver – and although it was very sensitive, you could only use it to do radiometry, which involves measuring electromagnetic radiation at specific frequencies.

This is how the Rhodes/HartRAO SkyMap was born.

It all started with two Rhodes University lecturers: Prof Eddie Baart and Dr Gerhard de Jager. In 2003, Prof Baart had retired from Rhodes, but still lectured some courses. He taught a first-year course on measurement, and unknowingly put every student in that lecture theatre in their place. Now there are many lecturers who try to deflate the self-importance and air-consuming egos of first-year students, but this grizzled man with a mane of white hair made a group of students realise how little they knew about the world without even trying.

Using simple science and logic, with none of the malice that coloured the agendas of other lecturers, Prof Baart made a group of gung-ho first-years realise their insignificance, sandwiched somewhere between gigantic galaxies that would outlive us and tiny bacteria that were dying as he spoke.

He was also one of the people who started what eventually became known as the Rhodes/HartRAO SkyMap, a survey of the southern sky at 2326 MHz. It didn't start as such an ambitious plan, though. The initial master's and doctoral degrees that came out of the programme focused on a portion of the sky. As Dr Gaylard says: "So, there was this receiver, with George's radiometer, and it could measure the total intensity of whatever you looked at … The first project was mapping a bit of the sky in the Scorpius region. I was involved in that, and that got published as [HartRAO's] first paper."

"No matter how big the area they surveyed, they still saw wisps of radio emissions from the Milky Way, and they soon realised they had to survey further and further from the Milky

Way, until they realised they would have to survey the whole sky," Dr Nicolson says.

While the 13-centimetre receiver was obsolete for Nasa's requirements, it was still more sensitive than the receivers Australian radio telescopes had. "The big point," Dr Nicolson explains, "was that because the receiver was so sensitive, very faint emissions could be detected. They could see emissions far beyond the band of the Milky Way, emissions that extended literally across the whole sky."

Justin Jonas, the man who became South Africa's first emissary on our SKA journey, was one of the first Rhodes University students who worked on SkyMap at HartRAO, and he eventually strung all the different research results together into one comprehensive survey for his doctorate.

But these students didn't simply come to HartRAO to enjoy their turn driving the telescope. They also had to develop the computer systems and software to process all the data they were collecting.

Dr Nicolson cites the example of Peter Mountford, who surveyed the Magellanic clouds for his doctorate. There are two Magellanic clouds – the Large and Small Magellanic Clouds – which are dwarf galaxies, comprising several billion stars. They are small-fry galaxies in comparison to our Milky Way, which has 200 to 400 billion stars.

In order to observe and collect data on these dwarf galaxies, Dr Mountford developed a lot of the software required for controlling the antenna, finally liberating astronomers from radio knobs and paper tapes. With these skills, he subsequently made a very successful transition into developing electronic instrumentation in the mining industry, Dr Nicolson notes. "So there was a big advantage in collaborating with Rhodes because they didn't simply come and use the system, they contributed to the development of the observatory," he says.

The SkyMap also contributed to the world's knowledge about cosmic background radiation, which is considered a strong argument for the Big Bang. Permeating the universe, there is an almost uniform thermal radiation – at about 2.7 Kelvin, which has the same magnitude as a degree Celcius.[2] The radiation is believed to be left over from the early universe because once neutral atoms had formed they could no longer absorb thermal radiation. The SkyMap provided extensive information about foreground high-frequency cosmic radiation, which is the electromagnetic radiation coming from our galaxy, and you need to have measured the foreground radiation in order to distinguish it from background radiation. "[The SkyMap] continues to be referred to as the map of the Milky Way emission, which you have to subtract off," Dr Gaylard says.

When the CSIR restructured in 1988, HartRAO's affiliation with Rhodes as an external user also contributed to HartRAO becoming a national facility, administered by the Foundation for Research Development. Later, in 1999, under the new democratic dispensation, it became one of the first national facilities of the National Research Foundation, which is still responsible for it.

As the world changed, HartRAO found new uses for the buildings around it, but for the most part has remained much the same. Through the day, the hiss of the receiver coolers creates a constant background hum, with the sporadic sound of moving machinery as the dish is moved to track a different celestial object. About 50 metres from the dish, a small spruit runs past the hostel. Today, Nasa's former bunkhouse is still there, a fairly large rectangular building, now partitioned into small monastic cells that shoot off a corridor. In fact, the peace is only disturbed at night, when a swarm of mosquitoes from the spruit launches

---

2 A degree Celcius and a degree Kelvin have the same magnitude, but the zero point of the Kelvin scale starts at $-273°$ C. It is known as "absolute zero" and is the temperature at which all thermal motion stops. The zero point of the Celcius scale is $0°$ C. So $0°$ C = 273 K, and a bright and sunny $30°$ C day in Johannesburg is also a 303 K day. In Kelvin terms, the day doesn't sound as bright and sunny so much as scorching hot.

a coordinated attack and tries to exsanguinate you. In summer, it is not uncommon for sleep to be interrupted by a loud bang, and someone shouting, "Die, you blood-sucking bastards. Die!" Well, at least that is how radio astronomers were awoken when I was staying there.

HartRAO now has an outreach building, in what used to be the mess hall, to show school learners and confused adults what radio astronomy is all about. You can now drive on a tarred road to get there. But at four o'clock, when HartRAO staff leave, it is just you and the dish, the afternoon sun baking the long highveld grass. There is only the sound of a turning dish, although now it looks more substantial and doesn't need people to drive it.

> **The children throw the Sun into the sky**
> A long, long time ago, before white people came from over the seas, before the big Bantu men moved south, the First Bushmen walked the land. But it was not the land we know today – the land was cold, and the sky was dark.
>
> The Sun did not amble across the sky. The Sun was a man who lived and talked and walked on earth, whose life-giving light shone from under his armpit. But he was a selfish man, who kept his light for himself, using it to light his house and garden. When he lifted his arm, the Sun's light shone bright and strong, but when his arm was down the night fell like a curtain with a red glow emanating from the Sun's house.
>
> One day, the women were sitting outside, enjoying the Sun's bright light on their faces, and wishing that it would continue to shine. One mother said to the others, "Why don't we throw the Sun up into the sky? Then we will always have light and it will reach all of us." While the other women nodded eagerly, no one wanted to steal into the Sun's house and toss him into the sky.

But an old grandmother, her hair salted with years of wisdom, called some small boys to her, and whispered her plan. Sneaky, sneaky, the children crept towards the Sun's house, moving from shadow to shadow. The grandmother had warned them to make sure the Sun was not looking, and cautioned them to wait until he was asleep.

After hours of waiting, the shadows lengthened as the Sun lay down to sleep, and in the dusky red glow the children snuck into his house. They carefully pulled him outside by his arms and legs, even though he was hot to the touch, and threw him into the sky.

And they shouted to him: "O, grandfather Sun, stay in that place, so that the Bushman rice may dry and you will make the whole world light."

That is why the Sun now shines its armpit light on the land. But the Sun does not speak – either because he is too far away for us to hear, or because he is still angry at the children of those First Bushmen.

3
The history of astronomy in South Africa

For the first European explorers, the Cape was a mixed blessing: a curse and a salvation. The seas were stormy and treacherous, and many ships met a splintery end on the submerged rocks, but it was a useful pit stop between Europe and the East. It was the turning point in the journey, when you began travelling more eastward than south. Portuguese explorer Bartolomeu Dias was the first European to set foot in the Cape, although when he initially tried to land there the weather was so bad that it blew him out to sea and he missed it entirely. He only managed to get there on his return journey in May 1488. He called the place the Cabo das Tormentas, or Cape of Storms. His king, John II of Portugal, later decided it needed a jollier name, and dubbed it the Cabo da Boa Esperança, or the Cape of Good

Hope, because it opened up a new trade route.

Now, being an explorer in those days was a rather hit-or-miss endeavour. There was no GPS, no maps, only the stars, a compass and luck. A lot of luck. But for European explorers, the stars in the Southern Hemisphere were a script that no one really knew how to read because, well, Europeans hadn't been that far south before.

The Dutch East India Company had a similar problem when it set up the Europe–Asia trade route in the 1600s, although by then Europeans had a slightly better idea of where they were going. But still there was no actual map of the stars to guide them. Dutch administrator Jan van Riebeeck came to the Cape of Good Hope in 1652 with the aim of setting up a trade stop for the company's vessels. In their book *The Astronomy of South Africa*,[1] Patrick Moore and Pete Collins make an important observation: for all the glamour now associated with the Cape, its flora, fauna and picturesque beauty, in the early days, it was seen as the land of milk and honey simply because it had Vitamin C or at least allowed sailors the possibility of getting some. Months on board ship with no fresh fruit or vegetables meant scurvy, the bane of sailors. You don't hear much about scurvy these days – whose symptoms include rotting gums, loss of teeth, even death – but in the seventeenth century it was as common as lice. It didn't help that no one knew what Vitamin C was, just that if you ate

---

1 This is the definitive book on astronomy in South Africa up until the 1970s. No detail is too insignificant for the two authors, and the book is, in a word, comprehensive – although admittedly too comprehensive for someone with only a passing interest in astronomy. This chapter relies heavily on *The Astronomy of South Africa* and the website of the South African Astronomical Observatory (SAAO) for its facts. If it weren't for these two sources, the history of South African astronomy would have long since fallen into obscurity and hearsay, and not enough gratitude and acknowledgement can be lain at the door of Patrick Moore and Pete Collins for their painstaking attention to detail. The book has been out of print for a number of years, but can be found at second-hand bookstores or in the SAAO's extensive library.

fresh fruit and vegetables – rather than the cured meat and grain found on sailing vessels – then you didn't get it. So one of the main aims of Van Riebeeck's station in the Cape was scurvy prevention, with supplies of fruity goodness.

The move provided a base for European astronomers who were sent to this unfriendly terrain to map the stars, mainly for navigation. The first of these astronomers was a French Jesuit priest, Guy Tachard, whose thick and impressive beard is also worth a mention. He stopped over in the Cape on his way to Siam, and left in his wake some brief observations on the "Croziers", otherwise known as the Southern Cross, and the fact that the European predictions regarding where the stars in the Southern Hemisphere should be were wrong. While his contribution may have been small – substantially smaller than many of the Cape astronomers who followed – Tachard is distinguished in being the first to set up an observatory, albeit temporarily, in South Africa.

The Cape's next astronomer – the Abbé Nicolas Louis de La Caille – brought new meaning to the word fastidious, and laid the groundwork for most of the astronomy that was to follow. In his two years at the Dutch trading station from 1751 to 1753, he charted the positions of nearly 10 000 stars. That might not sound like a lot today, when you can set a telescope to automatically scan the sky while you make a cup of tea, or take a photo and count the dots. But in those days, it was tough going. A word on what it was like to be an astronomer in the eighteenth century: tedious. You lived by night and, with painstaking attention to detail, noted the positions of all the stars you could see, as well as the time that you saw them. De La Caille's notebooks resemble accounting ledgers, with columns of neatly noted figures, times and positions. And those are just the observations – from there these positions have to be analysed to figure out the star's behaviour over time. It is a debatable point whether most modern people – in a world of texts, television and songs that last three minutes rather than overtures that went on for hours – would have the patience for such an exercise. Also, De La Caille was using a 71-centimetre telescope with an aperture of just under 1.5 centimetres. The Southern African Large Telescope (SALT) has an aperture of 9.2 metres, which makes it more than 600 times more sensitive than the equipment used by the French astronomer.

De La Caille, however, was also famous for a mistake resulting from measurements that were unfortunately taken in the wrong place. In the 1700s, scientists knew that in the Northern Hemisphere the Earth wasn't a perfect sphere – it was squashed a bit at the pole. And while they suspected that the Southern Pole mirrored the North Pole, they weren't certain. De La Caille upset many astronomers in Europe when his measurements found that the South Pole wasn't the same as the North Pole, in fact it was slightly extended, rendering the Earth pear-shaped. It took another 80-or-so years before Thomas Maclear redid the experiments, which had previously been thrown off by nearby mountains.

We take so much of what we know today for granted: the fact that the Sun is the centre of our solar system and an oblate Earth is one of many planets orbiting it, that you get Vitamin C from fruit, that malaria[2] is transmitted by mosquitoes. But some poor fastidious scientist spent a fair portion of his or her life figuring these things out, and 50 years from now – with the Square Kilometre Array, the Large Hadron Collider particle accelerator and genetic research, as well as thousands of other instruments and disciplines – our children will wonder how it was possible for us to be so ignorant, because the scientific leaps they take for granted will seem so obvious to them.

☾

In 1820, the British came to Africa, with flamboyant red army outfits,[3] stiff upper lips and an astronomer. As a seafaring nation,

---

2 The word "malaria" comes from the Latin *mal* (bad) and *aria* (air).
3 Following the final defeat of Napoleon in 1815, British army gear got

Britain understood the importance of star navigation, and since they'd decided to set up shop on one of the most inhospitable coastlines in the world, they knew they needed an observatory.

Fearon Fallows, the first head of the Royal Cape Observatory, graduated from Cambridge, and began what later became a trend – six out of the nine Cape astronomers were from Cambridge. Despite these academic credentials, he had no practical astronomy experience, and so spent the months before he sailed for the Cape on an intensive astronomy crash course. And yet he was the man who decided where the observatory should be. The Cape Town suburb affectionately known as "Obs" takes its name from southern Africa's first observatory, which is about 6.5 kilometres east of the inner city. But back in 1820, the outlying area of the settlement was an inhospitable swamp and the observatory was built on an island in its midst, an island known locally as Slangkop, or Snake Hill.

The observatory had to be located in the Cape – distant areas further inland were virgin and uncharted and it would have been a big ask to force an astronomer, lugging heavy and ungainly equipment, to embark on a mission into the hinterland, although a later astronomer did just that. But the Cape, even today, is known for inclement cloudy weather in the vicinity of its famed Table Mountain and for wind that can result in even a rather hefty woman clinging to a lamppost to avoid being blown away. In the old days, there were no buildings to shield telescopes from the sand whipped up by the wind, so the swamp was ideal. The sandstorms missed it and the dust that managed to inveigle its way into the area was damped by the marshy land. The Slangkop

---

increasingly flamboyant. But when the British had to defend their colonies against other colonial powers and the people who lived there first, red was not the most practical colour, especially if you wanted to be inconspicuous in the African veld. It took a number of defeats before they realised that feathered helmets and bright red togs didn't help you to blend into your surroundings. In fact, the Second South African War, with its prevalence of guerilla warfare, is noted as a turning point in the British Army's fashion sense. It's much easier to aim at a bright red target.

observatory was also in view of the sea, so Fallows could send signals and weather reports to the ships in the bay.

It took about a year to decide on the final site, a choice that gave later astronomers a name to curse under their breath as they fended off snakes and other natural trespassers, and then another eight years to build it. Fallows only had three years in residence before he died of scarlet fever, although many say that his condition was exacerbated by overwork. He is buried on the site – which is now the headquarters of the South African Astronomical Observatory – about 12 feet underground, allegedly to discourage grave robbers.

So that is where it all started: in a swamp. A number of notable – and other less notable – astronomers followed. Fallows' successor Thomas Henderson was notable in that he *almost* made a globally important discovery and managed to make observations of between 5000 and 6000 stars in his 13 months at the observatory. But from the sound of it, everyone who knew him in the Cape wished they could forget him. He hated the place, dubbing it a "dismal swamp", and the people who lived there hated him right back. But even today South Africa is famous for having made the observations from which the star closest to Earth was discovered. That star is Alpha Centauri.[4] It

---

4 Alpha Centauri is the closest star system to us, at about 4.3 light years – so it would take light from those stars 4.3 years to reach Earth. It's a "star system" because it's a binary star, with two stars orbiting each other. A third part to this system and the closest star to the Sun, Proxima Centauri, was discovered much later – also at a South African observatory, but more about that later. Alpha Centauri is also one of the bright pointers in the southern sky used to indicate the direction south. You take the long line of the Southern Cross (two lines make up the cross, a long one and a short one) and follow that towards the horizon. There are then two "marker" stars: Alpha Centauri and Beta Centauri. You draw a straight line between the two, and then take the perpendicular line (the line at right angles to the one between the two) and extend it towards the horizon. Where this perpendicular line and the extended long line from the Southern Cross meet is south. Convoluted, I know, but that's how ancient mariners and early explorers knew which way was south.

seems to be a reflection of Henderson's disappointing time in the Cape that, although he had made the accurate measurements, he delayed analysing his data until he returned to England in 1833. There's definitely a moralistic tale in here about procrastinating, because he was beaten to it by Friedrich Bessel, who announced the discovery of Alpha Centauri in 1838.[5] Being notable for an *almost* discovery sounds like grasping at straws, but it was still impressive considering the conditions at the Cape and the dearth of equipment.

However, the person who put the Cape observatory on the science map – for a *real* discovery – was an Irishman by the name of Sir Thomas Maclear, although his knighthood came later. He took over the reins of the observatory in 1843, and aside from his brilliance and resourcefulness, what makes Maclear such a pivotal character was his ability to attract equally brilliant people to him, two people in particular. Thomas Bowler, who is now known as an iconic South African artist, first came to the Cape as Maclear's servant; and Sir John Herschel, who was effectively the first friend of the observatory. Herschel's name is often linked to South African astronomy but, although he had an active interest in astronomy and published astronomical papers throughout his life, he was never actually part of the Royal Cape Observatory. His father, Sir William Herschel, discovered Uranus and was the first person to map portions of the Northern Hemisphere sky, so it would appear that the astronomical interest was hereditary. William Herschel's son followed in his footsteps, and extended his father's Northern Hemisphere map, and came to the Cape with the intention to do the same in the south.

Herschel Junior is the kind of man who would make anyone feel inferior, not because of his demeanour – he was known to be good-natured and friendly – but because it just doesn't seem fair

---

5 Some historians say that Henderson didn't publish his results because he thought them too significant, but, since this is a South African book, it must continue the reciprocal Henderson-South African tradition and be a mite ungenerous.

that one man could be so good at so many things. He discovered and named the moons of Uranus and Jupiter, was in love with his wife, made inroads into modern photography, had 12 children, was an active botanist, translated Homer's *Iliad* and was the president of the Royal Society.[6]

When he came to the Cape, he brought astronomical instruments in tow to undertake the herculean task of mapping the southern skies, although it must be noted – and not to detract from the seriousness and stellar quality of his work – that astronomy also provided a great diversion for his guests. "Tea and stars" was a common social activity for the Herschels' visitors.[7] And while Herschel was never actually part of the Cape observatory, his friendship with Maclear was invaluable to both. Herschel advised Maclear and provided peanut-gallery commentary as to how the new Cape Astronomer should tackle the problem that had been plaguing European scientists ever since De La Caille had taken meridian arc measurements in the Cape: how could the Earth be shaped like a pear?

Meridian arc measurements involve calculating the distance between two points on the same line of longitude. Northern Hemisphere scientists knew the distance from the equator to the North Pole, and it upset them greatly that the distance to the South Pole was allegedly different. Investigating this was one of Maclear's first tasks, made more pressing by the fact that the British astronomy mothership had even sent him equipment for testing Abbé de La Caille's measurements. The welfare of the instrument was also Maclear's responsibility, which may seem unimportant except for the fact that the instrument was about as high as a door, with a tent and protective gear, and had to be

---

6 This is technically the Royal Society of London for Improving Natural Knowledge, and is alleged to be the oldest science society in the world. It also happens to have a great motto: "Nullius in verba", which means "Take nobody's word for it".

7 Moore and Collins write extensively about "tea and stars" in *The Astronomy of South Africa*. Sounds like a great substitute for parlour talk.

lugged as far afield as the Orange River – by ox-wagon, in a place with no roads. It is not surprising, then, that it took him 10 years, although for scientists around the world there is no doubt that their relief – to discover that the Earth was a geoid, and that they hadn't been wrong for centuries – was definitely worth 10 years of another man's life.

After an illustrious career, a town in the Eastern Cape named after him and 36 years as head of the Cape observatory – for a catalogue of his many achievements, *The Astronomy of South Africa* is a must-read – Maclear retired in 1870, and when he died in 1876, he was also buried on the observatory grounds, not far from Fearon Fallows.

But the man who transformed the observatory into a world-class facility was David Gill, an austere-looking Scotsman – although descriptions paint him as a fairly affable fellow – who ascended to the helm in 1879. From childhood, Gill had wanted to be an astronomer, but his father was a watchmaker and wanted his son to follow in his footsteps. So, while wearing a watchmaker's hat during the day, he moonlighted as an amateur astronomer at night, building a telescope in his father's backyard and making his own observations. But it was a photo that changed the course of his life. Back in the 1860s, photography was still a nascent procedure.[8] Then, in 1869, 23-year-old Gill took a photo of the moon that made everyone sit up and notice the young watchmaker, including his future patron Lord Lindsay. Needless to say, he didn't remain a full-time watchmaker for much longer, although it was a decade before he was offered – and accepted – the post as Cape Astronomer.

In 1882, there was a singular event that opened a new avenue for astronomical research and furthered Gill's photographic reputation – a comet so bright that it could be seen with the naked eye in daylight. It was particularly spectacular from the Cape, with its relative lack of light interference, so Gill decided

---

8 Herschel was in fact the person who coined the term "photography" as well as a "negative".

to take a long-exposure photograph of it with a portrait camera to send back to the Imperial mothership. While the picture itself was impressive, the thing about the photograph that astounded people was the background, with its blanket of bright stars. This was the starting point of an endeavour that has made Gill a giant in South African astronomy and won him international acclaim: the *Cape Photographic Durchmusterung*.[9] He decided to photograph the southern sky and, using this technique, he catalogued nearly half a million stars in the night sky. At about the same time, in 1887, the Paris Observatory was on a similar photographic bent, and began the *carte du ciel*, or "map of the sky", which aimed to photograph the whole sky and pinpoint all stars above a certain magnitude.

It's called "astrography", and Gill is remembered as one of the pioneers of the field. Fortuitously, his brilliance in this nascent field caught the imagination of the international community, which was eager to beef up his arsenal of equipment. So when Gill retired in 1907, his successor inherited one of the best-equipped and modern observatories in the world.

Now something you should know is that scientists are sneaky. Astronomers sitting in the Cape were constantly pitting themselves against the adverse weather – the weather in Cape Town is a bit like the little girl with a curl in the middle of her forehead: when it's good, it's very, very good, and when it's bad,

---

9 *Durchmusterung* is a German word for a laborious and extensive survey of objects or data. In astronomy, *Durchmusterung* usually refers to the *Bonner Durchmusterung*, a catalogue of the Northern stars by Bonn University in Germany. The first version was undertaken from 1852 to 1859, and includes the positions and apparent magnitudes of about 325 000 stars. This map, and its follow-ups, were the most complete pre-photographic star catalogues. Since the original *Durchmusterung* surveyed the northern sky and a small part of the southern sky, other catalogues were added to it. One of these was the *Cape Photographic Durchmusterung*.

it's horrid. For astronomers, the dry Highveld, an inland region in South Africa that is between 1500 and 2100 metres above sea level, is nirvana: clear, dry skies, just right for star-gazing. So in 1902 the South African Association for the Advancement of Science started lobbying the Transvaal government for a meteorological observatory, with clandestine intent to do some sky observing.[10] It was rather fortuitous that in this year the area, as a result of the Second South African War, had just become part of the British Empire. The new observatory was officially called the Government Meteorological Observatory because the government felt that the farmers of the Transvaal needed information about the weather more than investigations into the stars. However, if that was the case, then they shouldn't have hired an astronomer to head up the observatory, which was commonly known as the Transvaal Observatory. If you give an astronomer a big telescope, you really shouldn't be surprised that he's going to use it to look at the stars. And if you are surprised, then you should have no part in running a country.

Enter Robert Thorburn Ayton Innes, who was made director of the Transvaal Observatory in 1903. By day he was a meteorologist and by night an astronomer, although it is

---

10 The fact that this association was lobbying for an observatory in the Transvaal at that time is quite incredible. In 1902, the Afrikaner forces had just been trumped by the British in what was a very ugly and drawn-out skirmish. The Anglo-Zulu War, and the South African Wars involved any group large enough to form an army pitted against all the others, often dragging smaller groups, which had been minding their own business, into the fray. The Second South African War lasted from 1899 to 1902, and afterwards the Transvaal was finally annexed as a British colony. These wars are often overlooked, and apartheid seems to have erased them from the collective consciousness, but they have global importance for a number of reasons. Aside from the British finally deciding to ditch the bright red uniforms, the term "concentration camps" gained prominence during this war, when a British commander, Lord Herbert Kitchener, decided that Boer women and black Africans were important conduits of Boer intelligence and so herded them into concentration camps.

worth noting that despite his meteorological hat, in his time the Transvaal Observatory accrued a great number of instruments that were perfectly suited to astronomy. But everything changed when the Union of South Africa was formed in 1910, and unified the separate Cape, Natal, Transvaal and Orange Free State colonies. Innes is alleged to have said that he was the first meteorologist to be made an astronomer by a decree of Parliament. When the country was unified, meteorological services were taken over by the Union Meteorological Department, which was based in Pretoria, leaving the newly named Union Observatory to drop its guise as a meteorological organisation. The Union Observatory is internationally recognised as the place where Innes discovered the closest star to the Sun, Proxima Centauri, making South Africa and Innes famous amongst the astronomical community. Innes' particular area of interest was double stars, technically known as binary stars. The discovery could be seen as a culmination of other astronomical work done in South Africa. More than 50 years before, the second Cape Astronomer Thomas Henderson missed out on the discovery of what for many years was thought to be the closest star system, Alpha Centauri, and later Cape Astronomer David Gill had pioneered the use of astrography.[11]

Innes was suspicious that there might be more stars in the Alpha Centauri star system – astronomers already knew about two of them, but the director of the Union Observatory thought that there might be another, substantially smaller and invisible to the naked eye. So he started to take numerous photographs of the system, and used a blink-microscope to analyse the photographic plates, and in 1912 he found it: Proxima Centauri, 4.2 light years away. It's small in comparison to the other brighter objects around it, but this red dwarf star is equivalent in size to about 15 Earths diameter-wise, and about a seventh of the Sun, which is 108 times the size of Earth.

---

11 Astrography is a bit of a tongue-twister. To make life easier, it can also be called "photography of the stars".

There were other observatories in South Africa, although none as illustrious as the Union – which was named the Republic Observatory when South Africa became a republic in 1961[12] – and Cape observatories. One was the Boyden Observatory in Bloemfontein. Durban tried to maintain an observatory of its own, but failed because of lack of funds. The money that was promised never manifested, and the Natal Observatory spent most of its short existence in its death throes, desperately waiting for a life-giving injection of funds that ultimately never came. Pretoria's observatory – the Radcliffe Observatory – was never really South Africa's. It was a mirror of the one at the University of Oxford, which confusingly was also known as the Radcliffe Observatory and had been compromised because of increasing light pollution as the city of Oxford expanded. So the university decided to set up a sister observatory in South Africa. The building was erected and astronomers took up residence in 1939, but there was the pesky problem of the Second World War, which meant that the observatory was without a telescope for a number of years. The delay, however, extended well past the war, and Oxford's new Radcliffe Observatory near Pretoria only took possession of its centrepiece 74-inch telescope in 1948.

What people did not expect, however, was just how fast South African cities were growing – great for the country's economy, but very bad for optical telescopes. Astronomers at the time must have known this, and so the decision to move instruments from the Cape and Republic observatories should not have come as a big surprise – although the subsequent outrage seems to belie the perceived common sense of the decision. In 1970, the CSIR announced an agreement with the British Science Research Council to consolidate all the observatories, with headquarters at the Cape Observatory and an observing outpost in Sutherland, of present-day SALT fame. Many a strongly worded letter –

---

12 In Dutch, it was known as the Zuid-Afrikaansche Republiek, which is why international currency markets refer to the South African Rand as ZAR.

the astronomer equivalent of taking to the streets in protest – condemned the decision to dismantle such a long tradition of astronomy in these observatories, which would effectively become redundant once all the instruments had been moved out. The phrase "killing the observatory" seems to have been used fairly often, and in retrospect rightly so. With the move to Sutherland, local observatories with all their history have either become administrative headquarters, like the Cape Observatory, or been closed entirely, such as the Republic and Radcliffe observatories. You wish you could go back and tell these very upset astronomers what would happen in the twenty-first century – that the move to Sutherland created an astronomical hub that has made it one of the best areas for optical astronomy in the world. For many years the largest telescope in South Africa was 74 inches in diameter, but because of the infrastructure and expertise at the Sutherland site, South Africa became home to one of the largest optical telescopes in the Southern Hemisphere, SALT, which has a 9.2-metre diameter.

Being an astronomer is a rather peculiar job. It is very solitary, and the people who excel in it – such as the individuals mentioned in this chapter, and in the rest of this book for that matter – dedicate their lives to a curiosity about the universe. The milestones reached in South African astronomy have been the remarkable work of individual people, who through their brilliance and – it must be said – pig-headed tenacity created something out of nothing, world-class science and observatories out of a swamp on the tip of a far-flung continent. The "death" of the early observatories must have seemed like a personal affront to the life's work of many people, and it's debatable whether the fact that all this became leverage in South Africa's bid to host the SKA – the largest scientific instrument in the world – would have assuaged the trauma or been cold comfort.

The amaXhosa and their understanding of the night sky
Temba Eric Matomela

AmaXhosa people, like all the other communities of the world, found the evening sky fascinating. The night sky served as a big laboratory where every community could observe, learn and identify some stars and other heavenly bodies to tell time, seasons and other occurrences. It is therefore safe to say that the Xhosa community had also used these celestial bodies for their calendrical systems, agricultural cycles, and the celebration of various rituals and customs.

I prefer to call this indigenous astronomical knowledge even though it somewhat lacks the scientific part of astronomy. We have lost most of the knowledge on this subject, which regretfully went undocumented to the grave with our forefathers, but we should consolidate and preserve whatever we have so that our children and grandchildren can be fore-armed with their rich cultural history and conquer the future well informed.

The amaXhosa are a part of the Nguni people who predominantly occupy the Eastern Cape. This is the area where the former presidents Nelson Mandela and Thabo Mbeki come from. In this area the night sky is clear and enchanting. The amaXhosa people identify stars by their magnitude and the time and season of the year that they appear in the night sky. The appearances of celestial bodies help them mark the celebration of various rituals, customs and agricultural cycles. When looking at the night sky they could not differentiate between stars and planets, not until telescopes were invented!

The most keenly observed stars are the Pleiades, which are known as *isiLimela* in Xhosa and Zulu. This group of young, hot stars is always observed in the early morning sky during the month of June. In fact, June – called *inyanga ye-Silimela* in Xhosa – is named after these stars. When they appear, people know it is time to till the ground for ploughing. It also indicates the time for initiating boys into manhood, and the years of manhood are counted by these stars. June to June is equal to one *isilimela* (one manhood year). If a boy is initiated in June, the following June that initiate will be one *isilimela* old as a man.

In a way, these hot young stars symbolise a new beginning for an initiate. A young man who has just been initiated is not only perceived as a new person but is also regarded as spiritually connected to the stars and better able to communicate with the spiritual ancestors who roam among the stars.

The next group of stars important in the Xhosa community is Orion's Belt. In Xhosa these stars are called *amaKroza* (stars queuing in a line). They represent three sacred places: the Grave, where the bones of our ancestors are peacefully resting; Starry Heaven, where the spirits of our ancestors are roaming among the stars; and Third Heaven, where the supreme God (*Qamata*) dwells.

People are not allowed to indicate these stars with a pointed finger; rather, you bend your finger to show respect when pointing or referring to these sacred places. Failure to observe this custom will result in the culprit being punished by God and warts will grow all over his or her face.

Canopus is known as *uCanzibe*. This star is the harbinger of the winter season and the month of May is named after it. This is the month that precedes the month of initiation (June) and all the boys preparing for initiation would be mourning their boyhood by donning ragged clothes and wearing half a calabash on their heads with porcupine quills embedded on it. They then go through the village in a group, blowing a whistle called *uvithi* or *isaBhabha*. When they do this, everyone in the village will know that they are mourning their boyhood ahead of their initiation.

Venus is a well-known celestial body among the Xhosa people. Traditionally, they saw it as a star and followed it right through the day. When Venus is visible in the morning sky, it is known as *iKhwezi lokusa*, the morning star. It is associated with diligence as it appears early before the Sun rises. Children are even named after it. If the child is a boy, he will be named uKhwezi; if a girl, she will be named *unomaKhwezi* with the hope that that child will also be as diligent as Venus in doing his or her morning chores before the Sun rises.

When Venus is visible at midday next to the Sun, it is called *iKhwezi lesibini*, the second Venus. When boys are herding cattle in the fields, they may sometimes play a game and ask an unsuspecting boy to point to the second Venus. If the boy points in the wrong direction, he then has to run and find the cattle that have gone astray and bring them back into the fold. This is just one of many such games that are played by boys to test each others' knowledge about nature and the environment.

When Venus is visible after sunset, it is called *uCelizapho*, meaning "he who asks milk from the one who is milking the cows". This is usually the time when the boys milk the cows. Normally when the milk bucket is full, the boys bring the calves to suckle from the cows, but instead a boy may look around and, if there is no adult in sight, quickly shoot milk from the cow straight into his mouth. After that the boy would look up and see Venus in the sky and would say to it, "You will never get even a drop of milk from me" – hence the name *uCelizapholo*.

When it gets dark and Venus is visible in the evening sky, it is called *umaDingeni*, a dating star. Boys and girls were not allowed to walk together in the public, so the only time that they could meet would be when the girls were fetching water from the fountain or from the river. Usually the girls would delay fetching water until late so that they could secretly meet their boyfriends at that time. This meeting, however, was just so they could kiss and that was all, no intimacy whatsoever.

Unlike some Western cultures, the amaXhosa associate shooting stars with bad luck. It is believed that when people die they become either good or bad ancestors or good or bad spirits. When you see a shooting star, you have to quickly spit on the ground and say, "Let the bad luck pass me by for I am not the only one who saw it." People believed that a shooting star is a bad spirit being kicked out of the celestial sphere by the good spirits. The spirits of our good ancestors are believed to be roaming among the stars, guarding and protecting us from the evil spirits. Contemporary beliefs are, however, different regarding a shooting star; these days many children believe it is a lucky star – a Western belief that has now been assimilated into our culture, which is evidence that our culture, like many others, is not static but dynamic and changes with the times.

When the full Moon is visible in the night sky, it is time to de-worm the children. Early the following morning, all children would be given de-worming medicine, such as aloe juice or dry ground pumpkin seeds, which were very effective at removing worms. A general belief was that at the time of full Moon, the worms are collected in one place in the stomach, so this is a perfect time to de-worm.

Crescent Moon – if it is sighted during morning time and is still ascending – is indicative of wind or cold the following day. It is sometimes called an overslept Moon.

Even though these stories or traditional beliefs lack scientific grounding, it is important to introduce the new generation to their cultural and traditional beliefs where the night sky is concerned. These stories show us that the amaXhosa were keen observers of their environment. Through their indigenous knowledge, they could tell time, mark their calendrical system, observe agricultural cycles and determine times for celebration of different customs and rituals. It is always interesting to learn about and compare the knowledge of the past as we venture towards the exciting new world of science and technology.

4
Southern African Large Telescope

You have to be dedicated to get to SALT at night ... or punctual. Because once the sun goes down and purple twilight creeps through the valley, you have to drive up the winding path to the site, about 18 kilometres outside of Sutherland, in the dark. And you can't turn on your car lights, as a sign kindly informs you – only your hazard lights. The rhythmic flashes of orange throw the black chasm to your left into sharp relief, and while contemplating certain death, driving at 10 kilometres per hour, you'll wish you had been on time.

The possibility of seeing SALT in operation at night is worth the year or two that the drive there might take off your life. SALT is the largest optical telescope of its kind in the Southern Hemisphere, and – since it receives faint light signals from distant stars and galaxies – it operates at night, far enough away from the nearest town so that there is no light interference.

During the day, the unmissable edifice of SALT stands 30 metres high on top of a lone hill. At night, it's impossible to see your hand in front of your face, let alone a building. All there is in the world is a sky so filled with stars that you get vertigo when you look into the heavens. That salted sky makes you forget which direction is up as it swallows you into its expanse.

The building itself comprises a cylinder 25 metres in diameter, with a domed top, and what seems like a rather thick chimney flanking the building. The primary mirror, housed in the belly of the building, is like a curved reflective honeycomb, permanently angled away from the ground – it is 11 metres across, made up of 91 individual mirrors, each a 1.2-metre hexagon. The mirrors are the primary collectors of the light, which is then reflected to a secondary collection of mirrors at the top of the telescope. The four relatively small mirrors – the largest diameter is about 0.6 metres – are necessary because they correct interference produced by the primary mirror. Also, in the "chimney" next to the building, there is an instrument that checks all the mirrors are aligned with each other.

SALT is designed to observe light from the near ultraviolet to near infrared range – about 320 to 1700 nanometres in wavelength, with the visible light spectrum sandwiched in between. There are currently five segmented mirror telescopes in the world, although South Africa's is the only one in the Southern Hemisphere. The "Big Five", as SALT science director David Buckley calls them, are made up of a collection of smaller mirrors and are 10 metres in diameter. There are two in Hawaii, the Keck telescopes; the Hobby-Eberly Telescope, at the McDonald Observatory in Texas, is considered SALT's sister telescope, and was used as the basis for the SALT design; and the Spanish telescope in the Canary Islands, Grantecan (Gran Telescopio Canarias), which was the last to be built.

South Africa's large optical telescope is the brainchild of

a consortium of 13 institutions from South Africa, Germany, Poland, the United States, the United Kingdom, New Zealand and India, and was more than a decade in the making. The telescope itself cost about $20 million, but the price for the three first-generation instruments on it varies from $5 million up. It costs about R20 million a year to run, and this cost is divided among the members of the consortium, depending on the initial share that they contributed. While the first ground was broken in September 2000, the telescope only became "fully" operational in 2011.

When asked how long SALT has been operational, Dr Buckley chuckles. "Well, it depends what you mean by operational."

It is one thing to build a telescope – to take plans, and construct a larger likeness from cement and metal – but that doesn't necessarily mean it will do what you want it to. This is why after the cement-churning machines have left, the remainder of the construction debris has been swept off the paved walkway and the red ribbon cut, the commissioning begins. Dr Buckley says: "This is when you try to use the telescope as you would when you're doing real science on it. That period is meant to shake down the telescope and check that it is meeting the specifications it was designed for." It's like creating a toy that you believe to be indestructible, giving it to a child and telling it to do its worst.

☽

But SALT isn't a toy: it is very sophisticated infrastructure made of a collection of finicky instruments. So diagnosing a problem is like trying to find a needle in a field of haystacks, when the biggest farmer in the area has just bought the two neighbouring farms which have fields covered in haystacks.

Two main problems were discovered in the commissioning of SALT, and took nearly five years to fix. The first had to do with the telescope system itself: the image quality across the entire telescope's field of vision was variable, which is a problem when you are looking at faint objects that can't be detected with the eye, such as a lit stadium on Mars.

Dr Buckley emphasises that the telescope was still doing science, but at a "somewhat reduced capacity". Eventually they discovered that it came down to the mounting of optics. The secondary mirrors and their assembly were supplied by a company in France (SAGEM) which duly built them, but was not responsible for mounting them. They are the mirrors that correct for optical imperfections in the primary mirrors. Now, these four correcting mirrors have to be perfectly aligned, and academically that makes sense because the light has to reflect exactly, but you only really discover just how important it is when they aren't properly aligned. But, unfortunately, with delicate optics, such as the ones on SALT, you can't kick them until you figure out where it isn't working[1] – you have to work back through a labyrinth of possibilities to divine the problem.

Once the issue was discovered, an interface between the mirrors and the telescope had to be designed so that the secondary mirrors would be aligned. This was successful and finally installed in 2010.

The second problem was with the main instrument: the Robert Stobie Spectrograph, or RSS, named after the late Robert Stobie, a past South African Astronomical Observatory (SAAO) director and first chairman of the SALT board and one of the original instigators of SALT. The telescope collects the

---

1 It is a truth universally acknowledged by all – except most electronics experts or engineers – that you fix recalcitrant electronics by kicking them. It knocks the dislodged wires/mechanics back into their ordained position, or causes its innards to shift so far away from where they should be that you can see the actual problem. There is also the possibility that it instils the fear of God into your computer/gate motor/cellphone, and gives it a taste of what is to come if it really kicks the bucket. Also, taking electronics equipments for drives past junk yards has been known to help show them what happens to computers/gate motors/cellphones that don't want to do their job.

light, and then instruments are used to analyse the data. There were three first-generation instruments on the telescope. First-generation instruments are usually cutting-edge once-off designs tailored to a specific telescope, and this is why they are built and designed by astronomers and engineers, rather than industry. The misbehaving apparatus in question has a $5 million price tag and was designed and built by the University of Wisconsin-Madison in the United States, with Rutgers University and the SAAO responsible for certain key components.

In terms of astronomy, a spectrograph splits up the light coming from outer space into its component colours: red, orange, yellow, green, blue, indigo and violet. You can do the same thing with a prism, breaking up white light into seven colours. Spectroscopy, however, shows something interesting: bright or dark lines appear in what you would expect to be a continuous rainbow of colour. These are called spectral lines, and with this information an astronomer can deduce what frequencies are being absorbed or emitted by a celestial body. Every element in the Periodic Table of Elements produces unique spectral lines at particular frequencies.

German optician Joseph von Fraunhofer – whose portrait depicts him as a rather shifty-eyed young man with impressively sizeable lamb-chop sideburns – was the first person to study in detail the dark absorption lines in the Sun's spectrum, now known as Fraunhofer lines (see Figure 1 in the colour section). In 1817, he published the wavelength of the solar absorption lines, and the chemical to which they correspond.

For example, an absorption line in the green section of the spectrum corresponds with a form of iron, and sodium is seen strongly in the orange, showing that these elements are present in the Sun.

Spectroscopy is very important in astronomy. At the moment – although who's to say they won't be able to do it in the future – astronomers cannot travel to a different planet or star, capture a vial of its atmosphere or a test tube of its soil and analyse it.

They rely on spectroscopy to discern the chemical composition of distant objects. It also enables them to discover how fast objects are moving away; for example, how stars are moving in our galaxy, whether there is an unobservable black hole at its centre or if there are exoplanets around certain stars, all through the Doppler Effect.

The Doppler Effect causes the frequency of a wave to alter as it moves. For a real-world-not-outer-space example, this effect causes the sound of a police siren to be at a higher pitch as it approaches you and at a lower pitch when in moves away from you. Both sound different to when you are standing next to the wailing and immobile police car. In astronomy terms, the Doppler Effect tells astronomers whether an object is moving away from or towards Earth because there is a shift in the absorption lines.

Needless to say, the SALT equipment that detects spectral lines is very high-tech. The RSS instrument was designed to be particularly effective in the blue region of the spectrum, or the shortwave area, but when the commissioning began astronomers were disheartened to discover that not only was it not performing to specifications, but that the quality of the data was skirting towards dreadful.

In order to detect spectral lines, the RSS is made up of 18 multiple lenses, clustered into groups. But it isn't as simple as a pair of binoculars, where you have lenses at each end and air in between. Some of the light hitting the glass is reflected – think of the glint of windows in the sun. As Dr Buckley says: "If you have normal air-glass reflections and you don't do anything about it, you'll lose about 4% of the light. Obviously (in astronomy) you want all the light to go through – 4% doesn't sound like a lot, but if you take 20 surfaces and you lose 4% on each, then at the end you've lost a lot of light – more than half of it, in fact."

There are several ways that astronomers ensure that their lenses transmit as much light as possible: by removing the nuisance of reflection, using very transparent and sometimes unusual materials for the lenses, and by inserting fluid between closely spaced lenses instead of air.

This comes with its own problems, of course; the first is cost. Many of the RSS's lenses are made of several crystalline materials: fused silica, otherwise known as silicon dioxide, which is a bit like glass; calcium fluoride, which is so fragile it will shatter if you look at it askance; and sodium chloride, which is actually table salt. You don't just walk into a hardware store or a grocer and order mammoth-sized crystals of pure salt in a very specific shape; it has to be specifically tailor-made to specifications.

Now many South Africans know about Sutherland. Aside from its infamy as an astronomy location, it is the town that, during the evening weather forecast, makes you feel better about your world simply because you don't live there. It has some of the most extreme temperatures in the country. In summer, you sit closer to the fan and thank your lucky stars that you're not in Sutherland because the mercury habitually loiters in the region of 40º C.[2] Similarly in winter, you draw your blanket up to your chin and radiate gratitude when the weather man tells you that Sutherland's maximum temperature won't breach the double-digit below freezing mark.

Extreme temperatures, however, give telescope designers a headache because, for example, lenses shrink and expand with temperature changes, by different amounts depending on what they are made of. Clearly just glueing such fragile lenses together would be a disaster, and a very quick way to fracture a great deal of money. So in order to compensate for both the bipolar temperatures and reduce light loss, fluid is used between closely spaced lenses. This fluid flows into the gap from an expansion bladder via a tube, so that when the lenses expand there is less

---

2 When these chapters were shown to people who spend a great deal of time in Sutherland, they – rightly or wrongly – accused the author of hyperbole: "Its altitude means it rarely gets hotter than mid-30s. Worcester is far hotter!" one protested. The author may grudgingly admit that they could possibly have a point. However, it is still too hot to want to be there.

liquid between them (because otherwise they would crack), and then when they contract fluid is sucked into the gap between the lenses.

With what Dr Buckley describes as "great dismay", undoubtedly a euphemism, the fluids that were used to couple the lenses unexpectedly began to chemically react with the actual materials that contained them. This resulted in the formation of polymers in the fluid, which effectively absorbed light – a bit unfortunate when you're trying to maximise your light throughput. Not only did this take time to diagnose, but then new materials and fluids had to be found and tested for compatibility.

To make matters worse, while the spectrograph was being disassembled, a technician "mishandled" one of these fragile calcium chlorine lenses (or looked at it in the wrong way) and it shattered.

The whole process took about two and a half years, with the RSS coming back to South Africa in 2009, and only being installed on SALT in 2011, following telescope modifications to improve quality.

Dr Buckley now laughs ruefully at a problem that must have caused no small amount of angst. "That just shows you the nature of the beast. When you're doing something for the first time, and you're pushing the boundaries and no one has done it before, then you find things out."

☾

SALT Lesson 1: Be on time.
SALT Lesson 2: Take a jacket.

A lift takes you to the second floor of the SALT building, and you step out into cold darkness, and are overcome by an expansive space. Even though it is early spring, an inquisitive cold October wind runs curious fingers across your face and finds ways to get inside your clothes. It takes a couple of seconds for your eyes to adjust – although that isn't saying much – and you can see the stars through the opening in the domed roof and the missing panels in the curved walls that enclose the large space. Hence SALT Lesson 2: take a thick jacket, because you will want to spend as much time there as possible, and chattering teeth are a distraction and difficult to disguise.

Seeing SALT in photographs doesn't prepare you for just how big it is. It is about three storeys of cold metal that moves like an enormous robot with an audible hiss of pressurised air, making anyone in its vicinity feel like Gulliver on his travels in the land of Brobdingnag, surrounded by giants. It moves on cushions of compressed air, and you can hear the cushions deflating as the giant contraption settles into a stationary position. All of this compounds the intense cold. There are gaps in the wall panels to cool the room, but the equipment itself is also chilled to remove the heat from the electronics, so it is also surrounded by a complicated knot of tubes, piping anti-freeze into the giant's veins.

The telescope is tilted at 37 degrees from the horizon to optimise the viewing window. This means that SALT sees a ring of sky. The closer the angle gets to the vertical line, the smaller this ring becomes. But the wider the angle, and the closer the telescope's eye gets to the horizon, the more interference there is from the atmosphere.

The atmosphere absorbs and scatters light, which means that less of the light can be seen by the telescope. The atmosphere is the reason that a sunset morphs into a tableau of colour. The light from the sun hits the atmosphere and dust particles in the air causing the light to appear red because red has the longest wavelength of the visible light spectrum.

So telescopes are optimised for maximum sky coverage, a balance between viewing area and atmospheric interference. To try to mitigate one of the problems caused by the atmosphere, SALT is fitted with an "atmospheric dispersion compensator" inside the eye of the telescope. "It compensates for the fact that the atmosphere acts as a prism, spreading out the light," says Dr Buckley. "You can't really see it by eye, unless you see a star low-low on the horizon and sometimes you'll see that it will change colour, going from red to blue and pink. What really happens in the atmosphere is that the blue light and the red light are bent slightly differently, so a star will end up being slightly elongated, bluer on one side and redder on the other. That compromises how sharp the images can be." The compensator, which optically corrects for this effect, is a set of prisms that can cancel the effects of atmospheric dispersion.

A quick aside: if you ever see published photographs of a telescope and its surrounding landscape, know that there was an argument involved. Astronomers are divided on the matter of clouds in photos of telescopes, simply because clouds are the anathema of optical astronomers. While radio astronomers can effectively see through clouds at most frequencies, white fluffy pillows of condensation can ruin an optical astronomer's night. Some astronomers believe that realism is important, and that clouds make the picture more aesthetically pleasing, but others believe them to be offensive. Such are the arguments of astronomers.

The hype around large scientific instruments makes people forget their purpose: they are built so that a small number of scientists can do very specific – and let's be honest, rather arcane – work. In SALT's control room, there are only three people: an astronomer, the telescope engineer and the occasional technical support guy or gal.

Because of the amount of custom work that goes into a telescope, each is a little like a frisky horse: it has quirks and can be somewhat temperamental. So SALT has dedicated telescope operators and engineers who know the best way to handle the telescope and its instruments. In terms of the division of labour, the operator negotiates the telescope, while the astronomer works with the instrumentation.

Only collaborating partners can use the telescope, and they send in proposals for reserving observing time. The amount of time allotted to a single institution is proportional to the amount of capital that that institution invested in SALT. So, for example, South Africa contributed about a third of the cost, so it is entitled to about 33% of the observing time. These proposals are adjudicated, with some accepted and others rejected. Then there is a dialogue, with much to-ing and fro-ing between the foreign astronomers and the SALT astronomers who undertake the observations, discussing how to best use the observing time.

SALT operates through an observing service, which means that SALT personnel do observing for all the astronomers using SALT, whether they're in South Africa or Poland. This makes sense for a number of reasons, mainly because if an astronomer were to fly out for the night, you would spend most of your time familiarising yourself with the equipment or scratching your head in bewilderment. It is much more efficient to let people who are well acquainted with the idiosyncrasies of the instruments do the observing. It also cuts down on costs, because no astronomy department has the budget to fly its astronomers to far-flung reaches of the world as a matter of course.

While the two telescope operators are based in Sutherland, SALT has six dedicated astronomers who spend most of their time in Cape Town, where the South African Astronomical Observatory's headquarters are. Roughly every six weeks, these astronomers spend one week on duty at SALT, depending on what observations have been booked and who is specialised in a certain area. This isn't a job for astrophysics graduates straight out of university: most of the SALT astronomers are on their second or third post-doc, and have experience working on other telescopes. It's a very specialised field.

Finding the object to be observed is made substantially easier through technology – when it is working properly, which is part of the reason why a computer engineer is part of the telescope technical support team, together with mechanical, optical and electronics experts. Also, SALT has a database of all the observing programmes and objects it observes and a specifically designed software program selects what is observable at any given time.

The only thing more high-tech in the near vicinity of the telescope and its technical apparatus is the coffee-maker: sleek, black, sexy and the focal point of the telescope's kitchen/lounge area. It isn't surprising, since being a night owl is an occupational necessity for people who work (literally) in optical telescopes. A word of warning: treat Astronomer Coffee, otherwise known as jet fuel for the body, with caution. One sip will have you up all night. A whole cup and you could be looking at a week of insomnia.

The control room is quiet except for the sound of music coming from the computer technician's laptop. Every now and again, someone calls out a song request – and sometimes negotiation ensues and song-choice veto rights are exercised. But for the most part the muted and rather sporadic conversation is about telescope focus and what is going to be looked at next, and the hours pass with the stars slowly moving across the sky.

☾

Spectroscopy doesn't give you the beautiful pictures you may expect from an optical telescope. You want to see the billowing multicoloured veins of the Crab Nebula, the glowing heart of a supernova, the tower sentinels of the Eagle Nebula's gas pillars when new stars are being formed. As someone who isn't a professional astronomer, you want to visually understand what is happening at the furthest reaches of the universe. A spectrograph, although interesting, just isn't the same – it's a bit like reading a synopsis of a book or seeing only the trailer for a movie.

Dr Buckley assures me that SALT takes pictures too, and even video. There are two other operational instruments on the telescope at present: the Salticam and the Berkeley Visible Imaging Tube (BVIT).

The Salticam is home-grown South African technology, developed and built by the South African Astronomical Observatory. It is a multipurpose instrument that includes a high-speed digital camera function. This means it can make videos of the stars. Like early motion pictures, it takes a number of still photographs at very short intervals, so that when the pictures are strung together it looks like a movie.

The BVIT's claim to fame is that it does very fast photometry of objects. Photometry is the measurement of the flux, or intensity, of the light. Light can be both a wave and a particle, called a photon. So while the telescope collects waves of varying wavelengths, this instrument can also detect the photons that are hitting the primary mirrors and where they are coming from, and make up an image. If a larger number of photons are detected, this corresponds with a brighter star. The fewer photons detected, the dimmer the object. The BVIT was installed on SALT in 2009, and commissioning took place over 17 nights in January and March. It is essentially a photon-counting detector system, which can compose images in microseconds – a microsecond is one millionth of a second. Unlike other similar devices, it creates no "noise" that would interfere with the detector and is capable of detecting photons in much shorter time intervals.

There are other telescopes on the Sutherland site, but the problem with being in the vicinity of such an impressive and renowned telescope is that they are literally and metaphorically in SALT's shadow. It's a bit like being unfairly cast in the role of ugly stepsister simply because you happen to be standing next to Cinderella. There are about 18 other telescopes on the site, as well as a magnetometer, which measures the strength and direction of the Earth's magnetic field, and a micro-gravimeter, which sounds like it belongs in a Batman comic. The superconducting micro-gravimeter is so sensitive that it even picks up audible noise, so it is encased in a cylinder and housed underground.

Many of these telescopes were transferred from Pretoria, Johannesburg and Cape Town. These major cities have observatories of their own, but as time passed their observations were increasingly affected by light and atmospheric pollution, so a number of observing telescopes were moved to Sutherland.

The great thing about having a collection of smaller telescopes, with a variety of collecting diameters, is that astronomers can do good science with them while others arm-wrestle for a slot on SALT. Case in point is the one-metre Elizabeth Telescope, named after the British monarch.

Early in 2012, a South African astronomer was part of an international team that found that there were planets around the stars in the Milky Way, and that this was the rule rather than the exception. "Wherever you look, you're likely to see a planet," says John Menzies, the South African involved in the collaboration.

Planet hunting has become increasingly important, as humans search for other planets that could support life. It may just be the most exotic career I've ever heard of: a planet hunter. Explorer meets astrophysicist. Swashbuckling, but with a telescope. A few decades ago, the idea of there being so many other planets was only found in the pages of science fiction, but in the last 16 years, more than 700 exoplanets have been discovered. These planets have all been relatively large, more like Jupiter than our tiny Earth.

Potentially habitable planets are difficult to find because not only are they small but they are also close to their parent star. These planets exist in what is called the Goldilocks Zone, in which it is neither too hot nor too cold. The reason that there is carbon-based life on Earth, the third planet from the Sun, is because it is the perfect distance from the Sun, and water can exist in liquid form. Mercury is the closest planet to the Sun and that means it's too hot for water to exist – it would simply evaporate because of the intense heat. Neptune, on the other hand, is the eighth planet from the Sun, and it is so cold there that the water is frozen.

However, this multiplanet discovery is a game changer, and it dramatically increases the chance of there being other

"Goldilocks" planets out there. The paper that was published was based on six years' observations, with 42 authors from around the world, and it all comes down to a difficult technique called microlensing.

Microlensing involves detecting planets through fluctuations in the brightness of a background star, and it is laborious work. Dr Menzies says: "It is a one-in-a-million chance that we'll see it. So we look at a million (stars) and one of them will be undergoing this phenomenon." The observations, which are still ongoing, take place in winter when the centre of the Milky Way is over the Southern Hemisphere, and Sutherland is not the only place where data is collected – astronomers in Australia and Chile are also using microlensing to discover planets. Because there are telescopes in Australia, South Africa and Chile looking at the same phenomenon, scientists have 24-hour coverage because something that is first seen in Australia will then be tracked by South Africa and then Chile.

Patricia Whitelock, acting director of the South African Astronomical Observatory and a world-regarded astronomer in her own right, gets very excited about these sorts of discoveries. "It's great to get such good use out of such a small telescope … and because it's small we were able to use a lot of time on it. You can't do that with the bigger ones.

"We've only recently discovered that there are planets out there. Everyone is looking for planets in the 'Goldilocks Zone.'" As she becomes more engaged in the subject, Dr Whitelock starts to speak faster, her British accent becoming more pronounced. "Life as we know it could exist on planets in the Goldilocks Zone because it is not too hot and not too cold," she says.

Dr Menzies says that you need a telescope in space to know the conditions on a planet. This is what Nasa's Kepler mission is designed to do: find habitable planets. According to the Kepler guide, it will "survey a portion of our region of the Milky Way galaxy to discover dozens of Earth-sized planets in or near the habitable zone and determine how many billions of stars in our galaxy have such planets". The Kepler satellite was launched in 2009, before the collaboration estimated that planets in the Milky Way were the rule rather than the exception.

The satellite weighs 1052 kilograms, and is equipped with only a photometer, the instrument that measures the intensity of photons emitted by stars in our portion of the galaxy. The data gathered by the photometer is transmitted back to Earth, and scientists spend hours, days, weeks analysing the information to detect regular dimming of a star's light. If the star's light dims periodically, it means that an exoplanet could be orbiting around it and blocking the light from reaching Kepler.

In December 2011, Kepler discovered two Earth-like planets – imaginatively named Kepler-20e and Kepler-20f – orbiting around a Sun-like star, Kepler-20. Now, before you get excited about the thought of taking a jaunt to one of these planets, remember that Kepler-20 is about 950 light years away. If you were in a car, driving at the 120-kilometres-per-hour speed limit, it would take you 8.5 billion years to get there. Some context: the Earth has only been in existence for about 4.5 billion years.

As time goes by, space is indeed being recognised as the new frontier. It is similar to the voyages of discovery when European monarchs sent off explorers to discover new continents: places that were completely new to them, that they hadn't imagined existed – although, obviously, not new to the people who already lived there. But the feeling of tangible excitement is the same, new vistas of opportunity. Space exploration now has the same flavour as those early voyages of discovery. This time, however, they really are new worlds.

5
Back to basics

What Karl Jansky discovered 80 years ago has become the foundation of a billion-dollar scientific discipline: radio astronomy. It all boils down to waves. Visible light, radio waves, gamma rays, they are all electromagnetic waves, but on different parts of the same spectrum, the electromagnetic spectrum. They are described by three things: their frequency, wavelength and energy.

The wave comprises a series of peaks and troughs, spaced at regular intervals, and is characterised by its wavelength, the physical distance from peak to peak. Radio waves occupy a large portion of the electromagnetic spectrum, from the extremely high frequency range, with a wavelength of one millimetre to one centimetre, to the extremely low frequency range where the

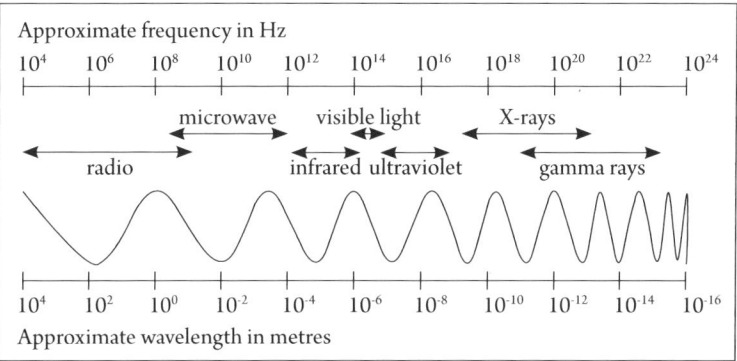

Wavelength

waves have wavelengths of about 10 000–100 000 metres. Light, on the other hand, has a much shorter wavelength of 380–760 nanometres, about 380 billionths of a metre. To put this into context, the shortest wavelength of light visible to the human eye – for the colour violet – is 380 nanometres. You could line up about 160 DNA chains, side by side, in that space. It's that small.

Gamma rays emitted from radioactive nuclei, which in a sufficient dose can cause deformation or death, are dangerous to humans because of their very small wavelength and high energy. The wavelength of light is so large relative to the distances between our molecules that the light bounces off us but with gamma rays, on the other hand, the radiation is able to penetrate our skin and interact with our bodies. Once you get smaller wavelengths than that, humans effectively become "invisible" to the radiation.

The wavelength measures a wave with respect to spatial distance, but scientists also want to calculate how waves behave with respect to time, and that's where frequency comes in. Frequency determines the number of times a wave peaks per second. So, if you have an electromagnetic wave with a long wavelength, there will be fewer peaks per second than a wave with a shorter wavelength.

While the underlying concept is simple, the machinery surrounding radio astronomy has become increasingly sophisticated and, well, expensive. In order to pick up and analyse celestial radio waves, you need three things: an antenna, a receiver and a data collector.

The most common radio telescope generally looks like a dish with a nose, much the same as a TV satellite. The dish is parabolic, like a bowl but not as deep. The radio signals hit the dish and bounce off it and, because it has a specifically designed curve, the signals collect in the nose, and bounce back towards the dish, aimed at a receiver in the middle of the dish. The receiver gathers signals with the frequencies for which the dish has been designed, and amplifies them before sending them to the data collector. This piece of computer hardware, the collector, transforms the analogue wave into a series of numbers. This is what the radio astronomer gets: the data signal information, combined with the specific time and location at which the signal was received.

It is worth noting that radio astronomy is not an immediate science. The image of an astronomer sitting in the telescope control rooms, watching beeps of data across a screen, noticing something interesting, shouting "Eureka!" and phoning everyone he knows to prepare them for his impending Nobel Science Prize nomination? Stuff like that only happens in the movies.

Astronomers have to wade through reams of data to find what they are looking for. Sometimes they only find a hint of it – whispering that something isn't as they expect – or a trace that they might be correct, but they need to listen and look harder. In order to do that, they have to do their experiment all over again, this time more precisely and honed to look for something in particular.

Now, that isn't as easy as it sounds. Time slots on radio telescopes, especially the large, sensitive ones, are booked months, sometimes years, in advance. If you think it's difficult to get a visa appointment or a specialist doctor's booking, it's nothing compared to getting observation time on a big telescope.

It also depends on what sort of astronomy you're doing. Optical astronomers work with light emitted by stars and galaxies millions of light years away, which is impossible to detect when the sun is shining overhead. That's why optical telescopes only operate at night, and their telescope operators are, by necessity, night owls. On the other hand, radio telescopes operate day and night, because the signals radio astronomers are looking for can be detected even though the sun is shining.

So it is often more difficult to commandeer time on an optical telescope than a radio telescope. But both instances require a lot of legwork and motivation.

First, astronomers have to submit a proposal to investigate a certain celestial phenomenon – which may or may not get accepted – and then they're allocated a time slot, which is hardly ever as long as they'd like. They send the telescope operator a detailed set of instructions about what to look at and how to go about it.

As one telescope operator once told me, astronomers from other institutions don't know the intricacies of a certain instrument as well as the operator. It's like when someone wants to borrow your car, and even though you've explained how the clutch sticks and told them the sweet words you have to whisper to the starter motor to get the engine to turn over, they never manage to get the smooth(ish) driving results you do.

This astronomer–operator divide causes no small amount of tension. The old idiom states that a workman always blames his tools. In astronomy, it can be the workman blaming another workman for how he uses tools that aren't his.

And then there's the fact that there's money involved. Big telescopes, radio and optical, are not built with pocket change. We're talking millions, sometimes billions, of dollars. In optical astronomy, the governments and institutions that built the instrument sometimes try to make back their money by charging for time slots, or the instruments can only be used by building partners.

It's a little different in radio astronomy. Up until recently, there was a "free skies" policy, mainly because the global radio astronomy community was so small. But now that it's getting more attention and the instruments are getting bigger, there's more money involved. Partner institutions want as many other players to buy into the construction and commissioning as possible, because that then dilutes the financial and risk burden. However, South Africa's MeerKAT will have a "free skies" policy.

Some people say that less is more, but in radio astronomy more is definitely more. A radio telescope is a huge receiver picking up radio waves from the universe, a rather large and expensive metal ear that is listening to what is out there. The bigger the receiving dish and the larger its diameter, the more signals it can detect.

So in an ideal world – devoid of radio frequency interference, where gravity is not an issue and science institutions have infinitely deep pockets – you could build a massive receiving dish, the Goliath of radio telescopes, so big that it would hurt your eyes to look at it. But the real world does not work like that. The larger the telescope, the quicker the costs rise like cream to the top of the limiting-factors list.

Also, it is more than a little bit impractical to build a gargantuan telescope. For example, the SKA will have a receiving area of one square kilometre, no surprise as to why they named it the Square Kilometre Array. If it were just one receiver, it would have to be more than a kilometre in diameter.

Radio dishes are made of different materials, with engineers preferring stronger, durable and lighter metals or composite materials, and naturally cost plays a factor in their choice. Take, for example, Australia's Parkes radio telescope, the second biggest single dish in the Southern Hemisphere. It is 64 metres in diameter, with a receiving area of 3216 square metres – the

SKA will be more than 300 times larger.[1] The Parkes telescope's weight, about 300 tonnes, or 300 000 kilograms, causes the parabolic dish to distort when it looks at certain parts of the sky, and necessitates some cunning science to account for and correct the distortion. Let's put this in perspective: a male African savanna elephant[2] weighs in at about 4000 kilograms. The Parkes telescope is the equivalent of 75 African elephants reclining rather uncomfortably on top of each other. In a very crude mathematical extrapolation, ignoring things like design, materials and the structures underneath, and a little bit of hand waving: if the SKA is 300 times larger, we are talking about 22 500 elephants. That's a lot of elephants.

This is where interferometry comes in. The technique uses a cluster of smaller telescopes, rather than one unwieldy leviathan. As the name suggests, interferometers are based on interference. A radio telescope measures the prevalence and direction of radio waves, with the machinery inside tailored to observe specific frequencies.

Now interferometry causes signals of the same frequency to interfere with each other, adding them together. This might sound rather counter-intuitive, but when the waves interfere, they become superimposed – one wave instead of two. Since astronomers generally focus on one celestial object at a time, the final wave signal gives astronomers consolidated data that they tease apart to unearth more information about the object.

The bottom line is that interferometry allows astronomers to look at the same object from slightly different places, as well as finding a way around the obvious problems of a Godzilla-sized

---

1 A quick note on units: 1000 square metres is not the same as 1 square kilometre. A squared kilometre is one million square metres – a kilometre (1000 metres) times a kilometre (1000 metres).

2 The African savanna elephant is the largest land animal. Some folk have grandiose size ambitions for Big Foot, or the Yeti, but the jury is out to lunch on that one. (No scientists were found on the jury.)

telescope. Also, it allows the telescope array to collect as much data as a very big dish.

If a star is very far away, the signal that reaches Earth is tiny, so many receivers are needed in order to amplify the information on that distant star. This is why, when you see pictures of radio astronomy dishes, they look like sunflowers moving in synch to face an unknown sun – all of them facing a star, or a quasar, or a planet, or whatever interests the astronomer behind the controls, so as to receive radio signals from it.

But, at the same time, not all radio frequencies can be observed. It's a bit like a corridor with many windows, some bigger than others. Walking along the passageway, you can only see the outside world at specific intervals or "windows of opportunity".

Technically, the lowest-frequency radio waves that can be picked up with a radio telescope are in the tens of Megahertz range, but these signals are usually reflected away from Earth by the ionosphere, which is effectively a charged layer of atmosphere that starts about a hundred kilometres away. These frequencies only reach the Earth at certain places, at certain times of the day and at certain times of the year, so your chances of picking them up are a bit hit-and-miss.

So ground-based radio astronomy can only really begin at the 30 MHz threshold, but most astronomers – those who enjoy the possibility of gainful employment all year round – generally deal with hundreds of Megahertz, most commonly starting in the thousands of Megahertz upwards.[3] For example, South Africa's radio astronomy observatory at Hartbeeshoek began life observing the 960 MHz band and upgraded to 2.3 GHz in 1964.

But there is a great deal of noise coming from the universe, which can drown out other frequencies, and in some cases it might as well be a double-insulated brick wall for all you can hear

---

3 1000 MHz equals 1 Gigahertz (GHz).

coming through. That also has to be balanced with interference from the atmosphere.

The range between 1 GHz and 10 GHz is, in the words of Goldilocks, "just right" – a balance between our own little bubble of atmosphere giving radio signals the cold shoulder and the cacophony of galactic sound.

Once you start hitting frequencies of 22 GHz, at the far end of the radio frequency corridor, you're met with brick walls on either side. At 22 GHz, where the signals have a wavelength of about 1.3 centimetres, water vapour blinds radio telescopes. In South Africa, where the rainy season generally falls in the summer months and the air is saturated with water vapour, there is little point in trying to observe celestial emissions at this frequency – although it's a work day for meteorologists measuring how water vapour exacerbates the greenhouse effect and causes temperatures to rise. During the dry winter months on South Africa's plateau, water vapour is less of a problem for dedicated radio astronomers.

There is a relaxation in frequency barricades until about 40 GHz, where oxygen becomes the next troll barring the gates, and by 60 GHz, oxygen blocks out all radio waves coming from the universe.

So, radio astronomy isn't as simple as setting up a dish, pointing it at the sky and waiting for the universe to unveil itself to you. Somewhere between the complicated machinery, arm-wrestling with other astronomers for observing time and finding astronomical windows of opportunity, it's a rather tricky business.

> **The origin of the Moon**
> Some say that |Kaggen, the Bushman creation and trickster god, threw an ostrich feather into the sky and it became the Moon; others say |Kaggen threw his red-dust-covered shoe into the heavens

and ordered it to shine. But one thing everyone agrees on is that the Sun hates the Moon. It may be because the Moon can speak, and the Sun does not. It could be that the Moon is lighting the darkness, which is supposed to be the Sun's job. For whatever reason, the Sun spends his days tracking the spoor of the Moon across the sky and, unfortunately for the Moon, the Sun is faster and stronger.

As the Sun rises in the morning, his sparkling knife in hand, he chases the Moon, managing to slice off only a sliver each day. As the days move into nights, the Moon gets thinner and thinner, until he is only a husk of his former obese self. And there in his empty belly, encased on one side by a shining silver backbone, rest the spirits of the dead.

When the Moon is emaciated and can run no more, he begs the Sun to spare his life and, although the Sun despises the Moon, he is not without pity and lets him recover his strength. So the Moon retreats to lick his wounds and fill up his empty stomach. And so it is that only when the Moon is strong, his stomach engorged and glowing, that the Sun begins his chase again.

6
The KAT-7 and the MeerKAT

A radio astronomer once told me that the best place to build a radio telescope was on the far side of the Moon. It is the quietest place humans can get to. Earth is incredibly noisy in terms of radio waves, with humans being the major culprit. Most of the things that we build emit radiation: cellphones, aeroplanes, microwaves, cars and – needless to say – radios and televisions. All these signals interfere with radio telescopes, and make it difficult to hear the relatively quiet sounds travelling light years through space. It is a bit like trying to listen to a song on your cellphone in the middle of a rock concert.

So when radio engineers are looking for a place to build telescopes, they try to find the quietest spot possible. Radio telescopes are so sensitive that the receivers inside the actual telescope have to be cryogenically cooled because even the receivers themselves emit radiation that interferes with the incoming signal.

While it is not possible to eliminate radio interference entirely, some places are quieter than others. The Southern Hemisphere, in general, is in a more advantageous position than the Northern Hemisphere simply because it is less densely populated and, geographically, because southern Africa, South America and Australia have landmasses surrounded by wide expanses of ocean.

The MeerKAT site in the Karoo was chosen for these reasons. The MeerKAT was supposed to be a small instrument to show the country's ability to build and operate a radio telescope, but the South African government decided that the country should build its own internationally competitive telescope and the MeerKAT grew into a 64-dish telescope. It will be built with taxpayers' funds, and will be the largest radio telescope in the Southern Hemisphere until the Square Kilometre Array comes online. The first five years of observing time have already been booked.

The closest town to the site is Carnarvon, which has a population of fewer than 2000 people and no industrial development. But even the town is a fair distance from the site itself. It could be easy to get lost in that part of the world – wide tracts of open land, a panoramic view of the horizon and a road that disappears into the distance ahead of you. There is, however, one important clue to finding the MeerKAT site: follow the swanky, new electricity poles.

The Northern Cape is one of South Africa's poorer provinces, and it shows. The government-built electricity poles that run alongside the faded-grey roads look like scarecrows that have fallen on hard times, lost their hats, clothes and balance. Two strands of cables connect the pathetic-looking wooden poles. But on the road out of Carnarvon, towards the MeerKAT site, they are watched over by the shiny sentinels of a new era: statuesque three-cabled poles, for electricity and data transfer, that connect the MeerKAT site with Cape Town. About 10 kilometres down the R63 road from Carnarvon, the robotic poles veer off to the right, like very expensive breadcrumbs leading you towards one of the biggest scientific projects in South Africa.

It is quite easy to find the project office – there aren't many buildings for it to compete with – a large, sprawling, colonial structure that cooks quietly in the Karoo heat on the Klerefontein farmstead. The farmhouse and a few other houses in the vicinity are occupied by the South African SKA Project. But the actual site itself is more than an hour away, along a dirt road flanked by the odd grizzled shrub.

There are few places in the world that feel as old as this part of the Northern Cape, and with good reason. About 250 million years ago, the Hantam region of the Karoo, where Carnarvon is, was a large inland sea. Travelling through it is like navigating your way along an ocean bed, the sandy expanse of the hills broken by seaweed-esque shrubs. It seems as though the only thing that has changed in all this time is that the water drained away – it is so isolated and quiet that you can easily believe you are a lone scuba diver at the bottom of the sea.

Unsurprisingly, you can drive 100 kilometres without seeing another car, person or animal, let alone a homestead. Unlike the Eastern Cape, with its verdant hills and the constant threat of livestock becoming road kill and transforming your vehicle into a tin can, there is an eerie lack of visible wildlife. It initially appeared to be a glorified desert, and so hot that any creature that lived there ran the risk of spontaneous combustion. But the creatures, albeit relatively small, are there. In fact, the Karoo is one of the most biologically diverse areas in the world; but in this challenging environment, with its scarce water resources and mean annual rainfall of about 20 centimetres, creatures are constantly on the lookout for predators and recognise that something as loud as your car probably means trouble.

All of this creates the perfect place to build a radio telescope: seismically, the area is very stable, it has a low annual rainfall and there's minimal radio frequency interference, with the only common activity being a tumbleweed moving lethargically with the aid of a slight breeze.

This is where the South African government bought the Losberg farm, slightly south of the middle of nowhere, and declared the area and its surrounds a National Astronomy Reserve, protected by some of the most radio astronomy-friendly legislation in the world.

But the problem with building a radio astronomy reserve is that the next government might change its mind and put a cellphone tower on the perimeter. This is where the Astronomy Geographic Advantage Act of 2007 comes in. The Act "[provides] for the preservation and protection of areas within the Republic that are uniquely suited for optical and radio astronomy". With this Act giving legal weight to her words, Science and Technology Minister Naledi Pandor declared the whole of the Northern Cape – with the exception of the Sol Plaatje municipality – an astronomy advantage area. What makes this different from a scientific organisation, or a sole politician on a mission, planting a flag in the ground and declaring an area a radio astronomy reserve, is that this area is now protected by law. Irrespective of political musical chairs, the MeerKAT site will remain a protected area, for the incredibly pragmatic reason that the only thing more difficult than getting something written into law in South Africa – with its extensive consultative process and public hearings – is getting something taken out of law.

So for time indeterminate, the area surrounding the Losberg farm will be an astronomy reserve. With its clear skies, small population and quiet surrounds, there are few places in the world so perfectly suited to radio astronomy.

☾

In 1994, when the African National Congress (ANC) came to power, many white South Africans were worried. Some chose to emigrate, predicting that the country would fall apart under the democratic dispensation; others stocked their pantries with tinned food and candles, suspecting an imminent descent into anarchy and chaos. Times were particularly uncertain for scientists, most of whom were white. Under the apartheid regime, South Africa became a world leader in a number of scientific fields, as a direct result of its military capabilities – to defend the white minority and to safeguard itself against external forces – and isolation from the rest of the world. For example, the weighty sanctions imposed by the international community meant the country needed energy security, which it found in biofuels, nuclear and coal-to-liquid technologies. In fact, Brazil – which is the world's second largest producer of bioethanol after the United States – uses South African technology developed in the 1960s.

Similarly, the majority of research and development in the country was channelled through either industry, which was the prerogative of private capital, or the military, which was the dominion of the government. In the 1950s up until the early 1960s, South Africa imported many of its armaments, until the United Nations Security Council established a voluntary arms embargo against apartheid South Africa in 1963, which was made mandatory in 1977 – and so, prompted by necessity, the state established a local arms industry. The data paints a picture of an increasingly militarised state: In 1961, the year that South Africa became a republic, official defence spending was estimated at R617 million – which in those days was a lot of money – and accounted for 7.7% of government expenditure. By 1989, these figures had burgeoned to R5.8 billion, and 15.2% of government expenditure.[1] It is quite possible that a great deal

---

1 P Bachelor, P Dunne, and D Saal, *Military Spending and Economic Growth in South Africa*, Centre for Conflict Resolution, University of Cape Town, in conjunction with Middlesex University Business School, 1999. It is necessary to have extensive footnotes when you write "military" and "South Africa" in the same sentence. It's a touchy subject, and the people

more money was actually spent, but through other conduits. In 1983, some 2% of South Africa's impressive defence budget was spent on research and development, which, although a bit behind the times, was the equivalent of Canada, Switzerland, Italy and Japan's spend on research and development in 1971.[2]

So, now the new government comes into power and it sits with a very real problem. The first thing it does is cut the defence budget, with spending being halved in real terms between 1989/90 and 1996/97.[3] If you are a new government in a very unequal society, you can't justify spending money on local defence – especially when the majority is now ruling the country, which means you would effectively be arming yourself against yourself – when that selfsame disenfranchised majority needs housing, education and basic necessities. And you also don't want to fund nuclear warheads posing as low-earth-orbit satellites.[4]

So scientists and engineers in parastatals had cause for concern – the sectors in which they were employed, which had been of vital importance 10 years previously, were no longer a priority. The defence budget had been cut, and the impetus to be self-sufficient in areas such as energy had evaporated; markets opened to South Africa and it was no longer a global pariah. South Africans could even travel internationally without

---

who are likely to take offence are also those who can track you down and enter into hostile and possibly physical negotiations about what you wrote.

2 S Langdren, *Embargo Disimplemented: South Africa's Military Industry*, Oxford University Press, 1989.

3 P Bachelor et al, *Military Spending and Economic Growth in South Africa*, 1999.

4 In the 1980s, the apartheid government build GreenSat, which it claimed was a low-earth-orbit satellite. But the satellite never got off the ground because sanctions meant no one in the international community was prepared to touch a South African "surveillance" satellite. So the government developed facilities to launch ballistic missiles. However, the ANC government took one look at a project that walked and talked like a nuclear weapons programme, and shelved it.

mumbling their country of origin under their breath and hoping they would be mistaken for an Australian.

But at the same time, countries need research and development to remain competitive, and require science and technology expertise to elevate the knowledge base. In a defence environment, scientists and engineers are paid to push the boundaries of imagination, and develop things that were thought to be impossible, while being safe in the knowledge that their job is secure.

As Dr George Nicolson says: "It was self-evident to almost everyone that when the [ANC] government came into power, their priorities would be housing, health, education, etc. – there was a huge backlog to catch up with. Would an ANC government be interested in basic research into the origins of the universe and things like that? It was a debatable point as to what extent they would support science."

The White Paper on Science and Technology of 1996 must have come as a pleasant surprise to scientists who were concerned about their future, and expecting the worst. The paper states: "Scientific endeavour is not purely utilitarian in its objectives and has important associated cultural and social values. It is also important to maintain a basic competence in 'flagship' sciences such as physics and astronomy for cultural reasons. Not to offer them would be to take a negative view of our future – the view that we are a second-class nation, chained forever to the treadmill of feeding and clothing ourselves."

Radio astronomy is a particularly benign science, with a large number of spin-off technologies. Large pure-science projects are one of the few places, aside from the military, where scientists and engineers can "play", with the knowledge of steady funding and a goal-orientated atmosphere. So rather than pumping money into weapons development, a radio telescope seems to be a better place to promote innovation through government funds – there is also the comforting fact that, unlike military technologies, it is unlikely to be used against you in the future.

The ANC-led government has spent more on radio astronomy than all the other South African governments since unification in 1910 combined.

Aside from the obvious need for a benign space to innovate, radio astronomy, with its use of cutting-edge technology and high skills requirements, speaks to modernity. South Africa's wealth is based on its rich mineral resources and, like many developing countries, it is seen as an exporter of raw commodities. By supporting astronomy, the country frames itself as a supplier of knowledge and innovation, and not the "second-class nation, chained forever to the treadmill of feeding and clothing ourselves" that the White Paper warns against.

It gives the country an international platform on which to engage with other state powers and attract foreign investors, not just to extract resources, but to engage and collaborate with its citizens as equals. Such an intangible goal helps to combat the equally obscure and yet pervasive perceptions that "Africans can't do technology" or are in some way inferior to more developed countries.

☾

By the sounds of it, South Africa stumbled into bidding to host the largest radio telescope in the world.

At this point it's necessary to introduce the man who is given the credit for making SALT a reality and beginning South Africa's SKA journey ... even if he won't accept the credit. The first thing Dr Khotso Mokhele says is "I don't like compliments", followed by "These people mustn't exaggerate". But as he speaks, a warm smile splits his face and you realise that this giant of a man may have a superpower – charm and sincerity radiate out of him, and you can understand how he managed to convince the international community to dig deep into their pockets for SALT. He says it was "an opportunity for a telescope to take South Africa into the big leagues".

"We succeeded in making the case that an instrument like that built in this country is more than just an instrument. It brings them into a project that allows them to contribute to what the project means to us," he says. For South Africa, SALT wasn't just about having a state-of-the-art telescope in the country – it was about being included in the international community in a highly skilled science, and enabling this to become part of the national identity. While the new ANC government struggled with addressing past inequalities, there was also a perception of African inferiority – which still persists today, especially overseas. "If the project of constructing the new society understood its real challenges in this country, that project would want to put South African funding into the telescope for a sense of ownership," Dr Mokhele says. By hosting something like SALT, the country assimilates into its identity the fact that, while there is widespread poverty, it is also the kind of country that can host a world-class astronomy facility. "We were able to bring a national consciousness into this project that makes them bigger than just fantastic science instruments," he says. "That is the beauty of these projects – they were driven by science imperatives, but at the same time they were driven by national imperatives."

But it takes a fair amount of time for these science plans to become a reality. Dr George Nicolson, who has been on astronomy committees since the 1960s, says that the first mention of a new telescope for the South African Astronomical Observatory was in 1986, and that design work began in the late 1990s. That means it took well over a decade to turn the initial sod. With such a long lead time for large projects, it isn't wise to rest on your laurels.

In February 2001, the local astronomy community held a national workshop to devise what to do next: ground had been broken in the construction of SALT, and South Africa needed a new project. Now, building a telescope is a laborious process, and not simply because of the construction and commissioning.

Before that even begins, there are meetings and convincing and fund-raising. "Just as we had secured the bulk of the funding for SALT, and understanding that with astronomy it takes 14 years from the first meeting to ground-breaking, I convened a meeting to bring senior astronomers together," Dr Mokhele says. "It took us 14 years to get to this point, what decisions should we take now so that 12 years from now we are breaking ground for the next project?"

Two main projects were brainstormed at that workshop. The primary project, the Virtual Astronomical Observatory, which will compile all astronomical observations on an Internet-based platform and is described as a "no-brainer", has slipped into the shadow of the SKA bid, primarily because at the time South Africa did not have the bandwidth capabilities, although with the increase in fibre optic cables this is becoming a reality. The second project was, of course, the SKA. The reason it was second on the list was because after forking out a great deal of money for SALT, which is 30% South African capital, there just wasn't money to spare on another large astronomy project, and it would have been difficult to justify asking for more funding.

Now, in the early days, South Africa didn't actually plan to bid to host the telescope. Initially, the country was just an observer, getting its foot in the SKA door, with Justin Jonas as its representative. Prof Jonas proposed the concept of the SKA at that meeting in 2001. "It was just a concept," says Dr Mokhele, rather than an ambition to host it here. "I said, 'Let's rather think about it as being joined up with this conversation that is taking place about this huge telescope.'"

Prof Jonas, who for many years held South Africa's SKA torch, isn't what you'd expect. He wears a short-sleeved collared shirt, with the SKA South Africa logo on the pocket, and a sensible pair of shoes. This soft-spoken man is modest about how he got chosen to pursue what has become the largest astronomy project in the history of the country, and you don't get the feeling that he would nominate himself. "I was the nominated person. In about

2001, I went to the first SKA steering committee meeting and I went to every one subsequent to that," he says in a quiet voice that you have to strain to hear.

Dr Nicolson is more effusive: "Justin was the obvious choice." Prof Jonas had been the person nominated to make the presentation on the SKA at the national astronomy workshop, and so was elected to pursue it; or, as Prof Jonas says, to "tootle off to meetings."

When he was not attending these SKA meetings, Prof Jonas was also the head of Rhodes University's Physics and Electronics Department, where he also took first-year tutorials. These tutorials were scheduled for 7.40am on a Thursday – which to a student is the crack of dawn, a time when no one should be alive, let alone awake. No one tells you when you're a student that people in the working world do that every morning. A possible reason for this omission is that realising that mornings get earlier and the work gets harder would disincentivise graduating ... or encourage everyone to become journalists, in which case public relations people can be shot for calling you before 9am. There was some suspicion among the students that these compulsory tutorials were set for that time in order to curtail their social lives: Wednesday night is infamous among undergraduate students, and especially first years, as a party night. So for the barely conscious and – let's be honest – frequently hungover students, it took all their energy to either stay conscious or concentrate; it was impossible to do both.

Prof Jonas would begin the tutorial by talking about equations of motion or classical mechanics, but would soon embark on an astronomical tangent, showing where his passion really lay. Careful attention was needed, or you'd find yourself holding onto the thread of a conversation you didn't understand. He radiated a quiet authority then, and he still does now as he explains to a very attentive, and not at all hungover, audience how South Africa progressed from a mere observer to a strong SKA candidate. "When we first started with the SKA, it was just with the intention of being involved in it. But fairly soon, within a year, I guess ... we thought we should be ambitious and consider being a host for the SKA," he says.

At that point, a number of countries were vying to host the gigantic telescope. The United States was considering Arizona as a possible site; later the Chinese also joined the conversation, offering a possible site. An Australian SKA site had always been a possibility because of the country's history and strength in radio astronomy. Also, they were very involved in the conceptualisation of the SKA because of their expertise and impressive credentials. But on the sidelines of these SKA meetings, astronomers were talking about South Africa as a possible candidate because of its outstanding climate and physical attributes.

A word about official meetings: the best and most interesting conversations almost always happen at the coffee table or in the smoking area. If you want to find out what's actually going on and learn the real intent behind the political posturing, go and strike up a conversation with one person – if it's a group, people are generally more circumspect – and then you'll find out what they really think, which doesn't necessarily correlate with their country's official position.

In that arena – the actual meeting, not the sideline conversations – Prof Jonas wasn't just a man with a mind full of radio astronomy; he was a South African emissary with an official mandate, and as interest in a South African site grew, he brought these calls back to South Africa. The powers that be – mainly in government, the National Research Foundation and the radio astronomy community – had a think about it and decided to put forward a bid. Dr Nicolson, Prof Jonas and Dr Mokhele describe the meeting with former Department of Science and Technology director-general Rob Adam[5] as one

---

5 Dr Adam was one of the initial drivers of South Africa's SKA bid, and when he left the Department of Science and Technology to join the South African Nuclear Energy Corporation (Necsa), he remained head of the country's SKA steering committee. He now works in the private sector and

of the shortest meetings they'd ever experienced. Booked for an hour slot, the meeting lasted for only 10 minutes before Dr Adam gave the project the go-ahead – they didn't even get as far as the Powerpoint presentation.

But you don't just raise your hand and hope that an international consortium with big money will pick you. You need to show that you are willing and able to host a mega-science project, and to do this you need to build an instrument. In South Africa's case, the reasons for this were twofold. On one hand, the country needed to demonstrate the viability of the site and the country's commitment to astronomy, and on the other it had to show that it had the capabilities and technical wherewithal to build and operate a radio telescope. The country had HartRAO, but that is like comparing a three-bedroom house in suburbia with the Palace of Versailles. Yes, they are both homes; yes, they are made from bricks with the odd window; but that is pretty much where the comparison ends. So, yes, South Africa had the Hartebeesthoek Radio Astronomy Observatory, but it needed something much more impressive to become a real bid contender.

The country had experience in lobbying to host an international science instrument, but this was different. "Quite early on, everyone understood how huge a science project this is; not just in terms of the size of the instrument, but huge in terms of the science question that could possibly be answered by this instrument," says Dr Mokhele. SALT was sold to partner countries as a great instrument in an ideal location; with the SKA, it started with the science questions: What happened after the Big Bang? How are galaxies formed? What is happening in the universe?

Everyone will tell you that South Africa's SKA bid only started to take shape when Bernie Fanaroff joined the project office. "The big watershed, when we got seriously involved in the site bid ... was when Bernie was brought on board," Prof Jonas says.

---

is still head of the steering committee.

There are many things people don't know about Bernie Fanaroff, like the fact that he, with his colleague from Cambridge, Julia Riley, published an article differentiating radio galaxies into two types, based on their shape. These are called Fanaroff-Riley galaxies, Type I and II, and that paper is one of the most cited in radio astronomy. After having studied at the University of Witwatersrand (Wits), Dr Fanaroff completed his doctorate at Cambridge in Astrophysics and returned to South Africa. Although he worked as a junior lecturer in Physics at Wits and started a research programme at HartRAO upon his return, he soon gave this up to join the trade unions in the fight for democracy.

But you would only know that if someone else told you, because Dr Fanaroff never talks about himself. If you try to broach the subject, he answers with a monosyllable and you can feel the silence fill the space between you. His work in the Presidency under former president and struggle icon Nelson Mandela has made him an expert at evading questions while being polite and diplomatic.

Whenever Dr Mokhele comments on Dr Fanaroff, his stock phrase is, "He is very polite." Dr Nicolson describes him as "self-effacing". "He plays his cards very close to his chest, and he will never openly come out with things he doesn't really want to broach."

So, getting Dr Fanaroff to talk about himself is nigh impossible. However, one thing everyone agrees on is that he was the natural choice to head South Africa's SKA bid.

Piecing different people's versions of the story together, co-opting Dr Fanaroff into the project went something like this: everyone in the science community knew – or at least knew of – Bernie Fanaroff. He had worked with Dr Nicolson at HartRAO before he abandoned radio astronomy to get involved in the trade unions, specifically the National Union of Metalworkers of South Africa. He had been director-general in the Presidency, was head of the Office for the Reconstruction and

Development Project (RDP) and the deputy director-general of the Department of Safety and Security. With a CV like that, it would be difficult for decision-makers not to know about him. So when Dr Adam and Dr Mokhele, as head of the National Research Foundation, cast their eyes about for a possible candidate, Dr Fanaroff was "an absolute natural", Dr Mokhele says. He was an astrophysicist by training, but also had management experience and impressive political credentials.

The importance of political credentials, although an issue often skirted over by most of the people interviewed, should not be undervalued. For a project like the SKA, you need complete political buy-in, which means that someone has to be able to explain the importance and relevance of such a project to the reigning powers. Radio astronomy, and highly specialised scientific equipment involving collaboration with developed countries, radiates elitism, rather than a pragmatic way to improve skills in the country. But 19 years working in a trade union, at a time when – euphemism warning – leftist leanings made your life very difficult in South Africa, and subsequent work in building a fledgling democracy, dilutes this impression. So combined with Dr Fanaroff's astronomy expertise and management experience, his ability to sell the project as something necessary and relevant to modern South Africa, with its myriad problems, cannot be underestimated. Not that he ever plays the "Africa is poor and needs help" card. As he says: "There's a dual message: that Africa requires the skills and capacity to become a major economic growth story, but also that it's seen as the next great business destination." He emphasises that you have to have people who can absorb the technology that the global economy wants to sell to Africa.

Dr Fanaroff was offered the position of project director in November 2002, and took office in January 2003. "Rob [Adam] wanted somebody who had some astronomy knowledge, but also had some knowledge of management and project management. That's why they asked me to do it," he says.

Listening to him speak, it appears that astronomy knowledge is secondary to management: "I focus more on the management, but general management rather than project management ... the issues surrounding the project and the interaction with stakeholders, both here and internationally." These "issues" centre on project communication, interaction with political stakeholders, the site bid process, things, he says, that "are not strictly speaking project management".

So in May 2003 South Africa, along with four other countries, submitted an expression of interest to be part of the SKA. Dr Fanaroff gives Sheereen Rawat, at the time a member of the SKA SA team, credit for pinpointing possible SKA bid sites. She identified a number of sites, which had to meet size, population, radio interference, seismicity and other criteria. But you can't measure these different things from a desk in Johannesburg. So members of the SKA Project Office, as well as some people from the Independent Communications Authority of South Africa, took to the road like hi-tech gypsies, with equipment built by Radio Frequency Interference Manager, Gerhard Petrick. Dr Nicolson describes it as "a big trailer with SKA and antennas and things that they could tow around" in search of the best radio astronomy spot – and one that would fit in with the SKA specifications. They visited 25 remote sites around the country before deciding on Losberg farm.

The African SKA has burgeoned from one radio astronomer "tootling off to meetings" into a project with headquarters in Johannesburg, an engineering office in Cape Town, a large farm housing millions of rands worth of equipment, and a staff complement that runs into the hundreds. From thinking that the country would just observe and get a foot in the door of one of the biggest science projects in the world, South Africa – undoubtedly the dark horse in the race – will have the most sensitive and impressive radio telescope in the world in its backyard.

Like Russian dolls, South Africa's XDM (eXperimental Development Model) and KAT-7 (Karoo Array Telescope) radio telescopes all grew out of the same idea: let's see if we can develop radio astronomy technology and test that it works. Initially, the plan was to build a 20-dish technology demonstrator for the SKA bid, which would be funded by the Department of Science and Technology.

Prof Justin Jonas explains: "It was initially going to be a very small telescope, just to show the feasibility of the site and the capability of South Africa. But then it was decided – and this was pretty much a top-down decision – that we should actually build something quite substantial, and that was the MeerKAT. We actually got funding because it was much bigger than KAT." But "much bigger" is a bit of a euphemism – it's David versus Goliath, a 10-dish array compared to what was originally envisioned as an 80-dish MeerKAT.

"The Department of Science and Technology applied to the Treasury to give us funding for a world-class telescope and to develop human capital. It had to be world class, or there was no point," Bernie Fanaroff says.

He describes how a large radio telescope got named after a small furry creature. "We wanted to build *meer* KAT," he says. *Meer* is Afrikaans, one of South Africa's 11 official languages, for "more" – more KAT. He says this is where the name "MeerKAT" comes from. However, a meerkat also happens to be an indigenous mongoose-like mammal, and there are a number of similarities between the – it must be said, very cute – furry animals and the MeerKAT radio telescope. First, the meerkat has a pointy nose that is rather similar to the nose of the MeerKAT dishes. Second, they move in packs, with a meerkat clan generally having about 20 members. But you sometimes find "super-families" of more than 50 meerkat – kind of like a super telescope array of 64 dishes. But for whatever reason, the name stuck and, with the necessary funding, Operation MeerKAT got underway.

The XDM prototype was meant to be a precursor for the MeerKAT, with design work beginning in 2006. The dish is 15 metres in diameter and only slightly less sexy than its KAT-7 progeny – that may be because they're smaller, with a 12-metre diameter, or because they have a daintier nose. The goal for the XDM was for South Africa to develop a radio telescope, which bloomed into the KAT-7, because you can't test an array of telescopes when you only have one.

This is an example of why having a tradition of military prowess comes in handy. "After 1994, a whole lot of military stuff stopped," says Dr Mike Gaylard, and yet we had many skilled people in the country who are trained engineers, technicians and the like. Many of them found a home in projects such as the SKA, but there were also whole companies that were at a bit of a loose end. One such company was IST, which in 2009 was bought by British arms manufacturer BAE Systems. "They built big fibreglass structures, buildings, factories and things like that ... somebody thought 'These guys could make a fibreglass dish,'" Dr Gaylard says. And that is what they did. On the site at Hartbeeshoek, a couple of metres from the 26-metre dish and right next to the XDM prototype, there appears to be a black flying saucer taking a quick break from exciting UFO enthusiasts.

Since IST is based in Pretoria, it made much more sense for them to create the first dish – Dr Gaylard describes it as pouring pancake batter into a rounded black pan – in Hartbeeshoek. First, it's closer than the Karoo; second, there are radio astronomers on hand to check that it actually works ... and to put out fires if something goes wrong. Engineering for radio astronomy purposes is different to other kinds of engineering, and explaining the intricacies of the science over the phone isn't as effective as having the brains on hand.

The great thing about making a mould for a dish is that it's fairly simple to make more. This is one of the KAT-7's claims to fame: it is the first array of composite fibreglass dishes in the

Construction of the tracking station began in January 1961. A photographer named Roelie van Wyk was employed to record progress, and he is caught in action here, hanging from the crane.

The official opening of the station in 1961. The antenna was originally built with an aluminium mesh surface, and operated at a frequency of 960 MHz.

The 26-metre HartRAO dish during an electrical storm in October 2002.

A panorama of HartRAO in September 2003. On the left is the Visitors Centre, with the hostel visible over its roof.

© M. Gaylard/HartRAO

Fifteen third-year physics students from the University of the North West visited HartRAO with lecturer Thebe Medupe for a radio astronomy practical in October 2002.

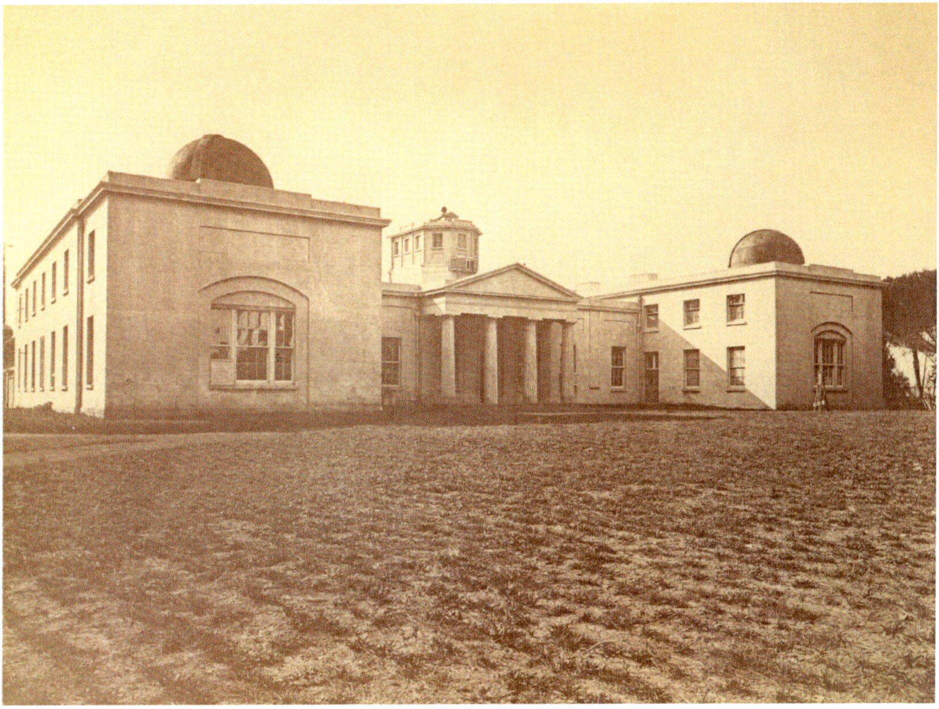

© South African Astronomical Observatory

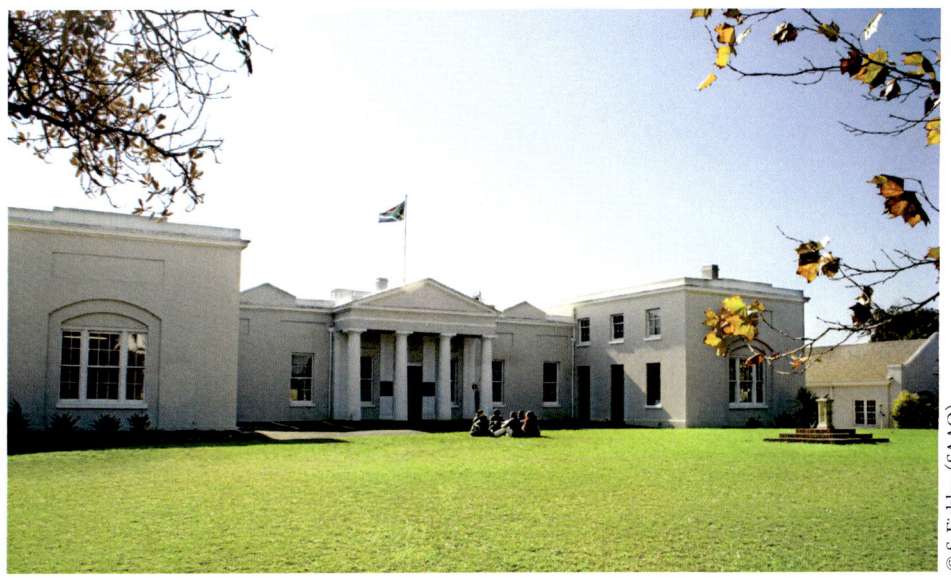

The Royal Observatory at the Cape of Good Hope, designed by John Rennie and completed in 1828.

© S. Fishley (SAAO)

The present-day South African Astronomical Observatory headquarters in Observatory, Cape Town.

This sign, and the ones the follow it at 10km intervals, are the only things that break the landscape surrounding the road to Sutherland.

A panoramic view of the Sutherland station with the Milky Way, Venus, Jupiter and zodiacal light in the background.

"SALT, sunset, Milky Way."

A fish-eye view into the heart of SALT and its primary mirrors.

An image of the Lagoon Nebula's central region taken by SALT. The nebula is a giant interstellar cloud – a soup of gas, dust and plasma – in the Sagittarius constellation.

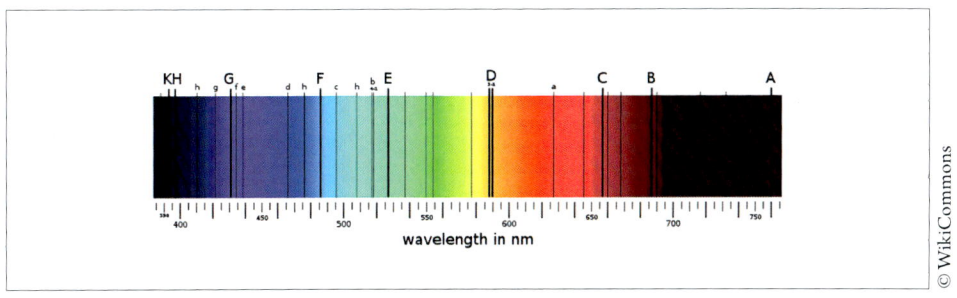

Figure 1: The solar spectrum with Fraunhofer lines.

The KAT-7 interferometer.

A close-up view of one of the KAT-7 dishes.

© R. Millenaar

© I. Heywood/SKA SA

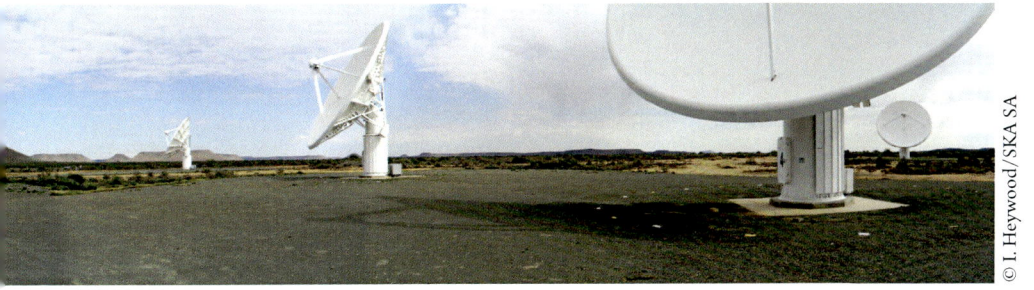

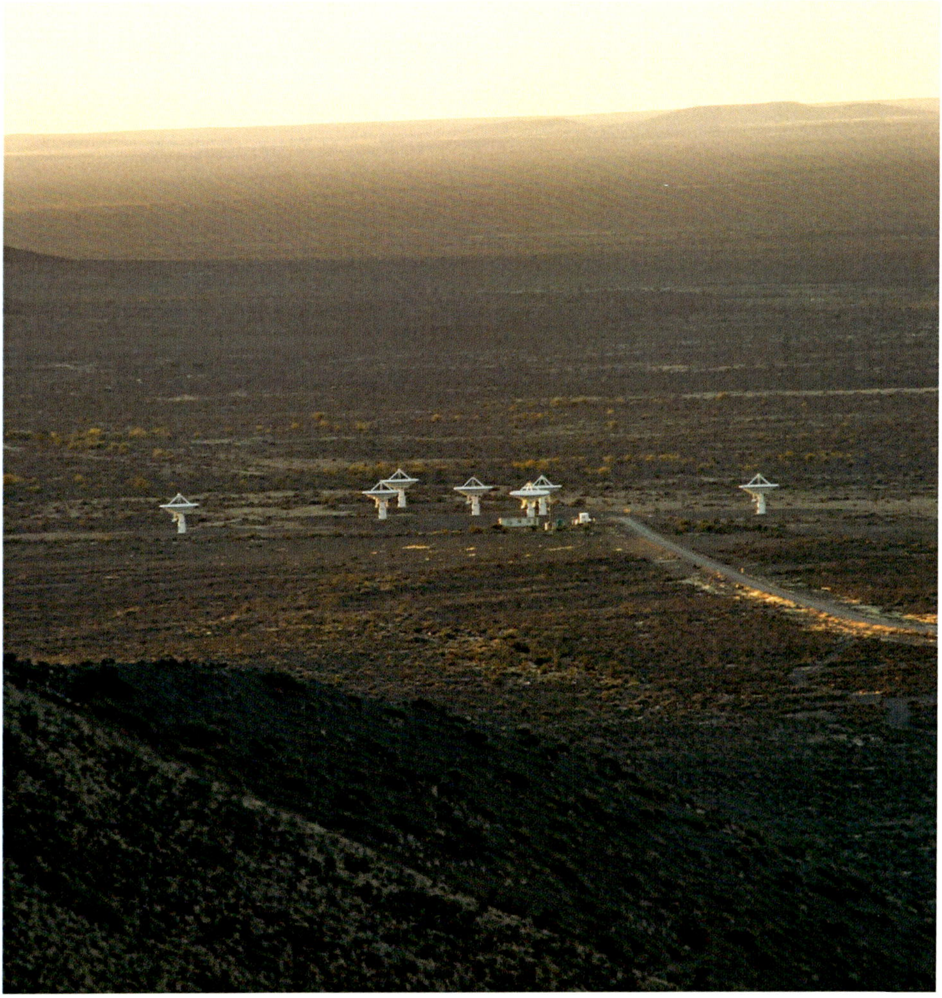

© N. Oozeer

Losberg farm and the KAT-7.

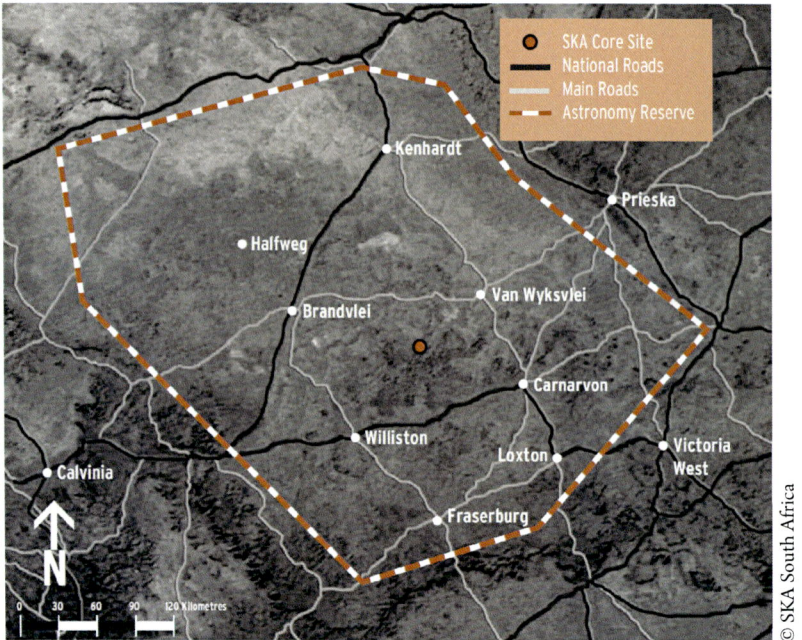

The Astronomy Geographic Advantage Act of 2007 "[provides] for the preservation and protection of areas within the Republic that are uniquely suited for optical and radio astronomy".

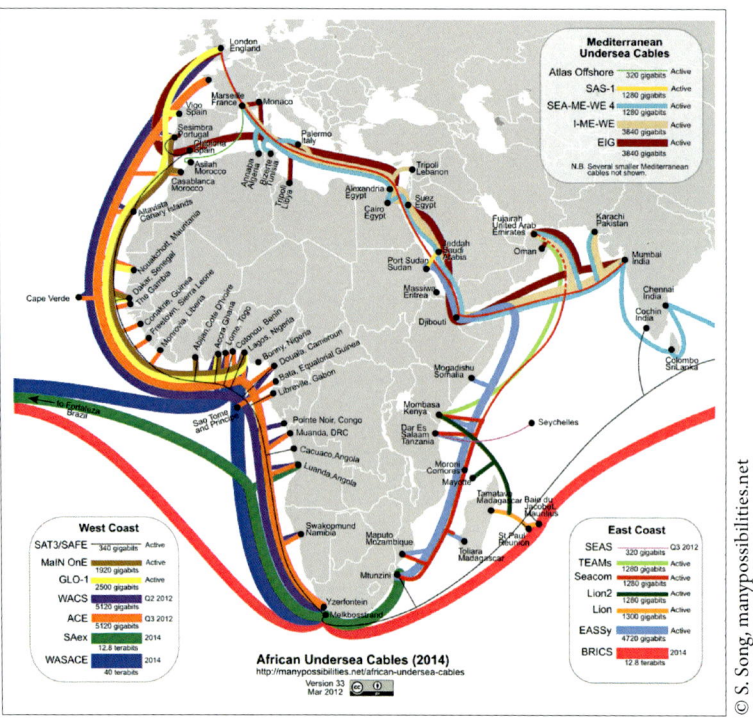

In recent years, Africa has received a glut of undersea cables. With more of the same to come, data transfer capacity may just catch up to the needs of the SKA, without any need to break the bank.

In 2012, Carnarvon is a one-horse town, but in 2019? Who knows.

© S. Wild

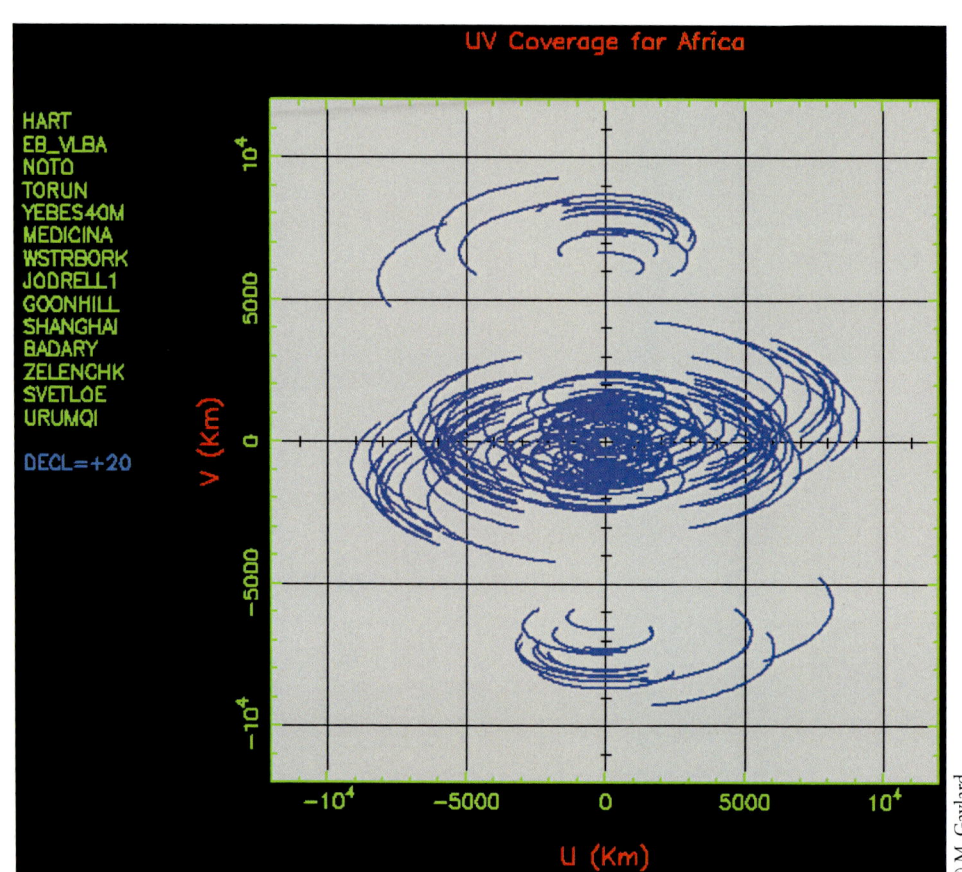

Figure 2: European VLBI Network without HartRAO produces inner tracks, adding HartRAO produces outer tracks. This is for mapping a radio source at 20 degrees North.

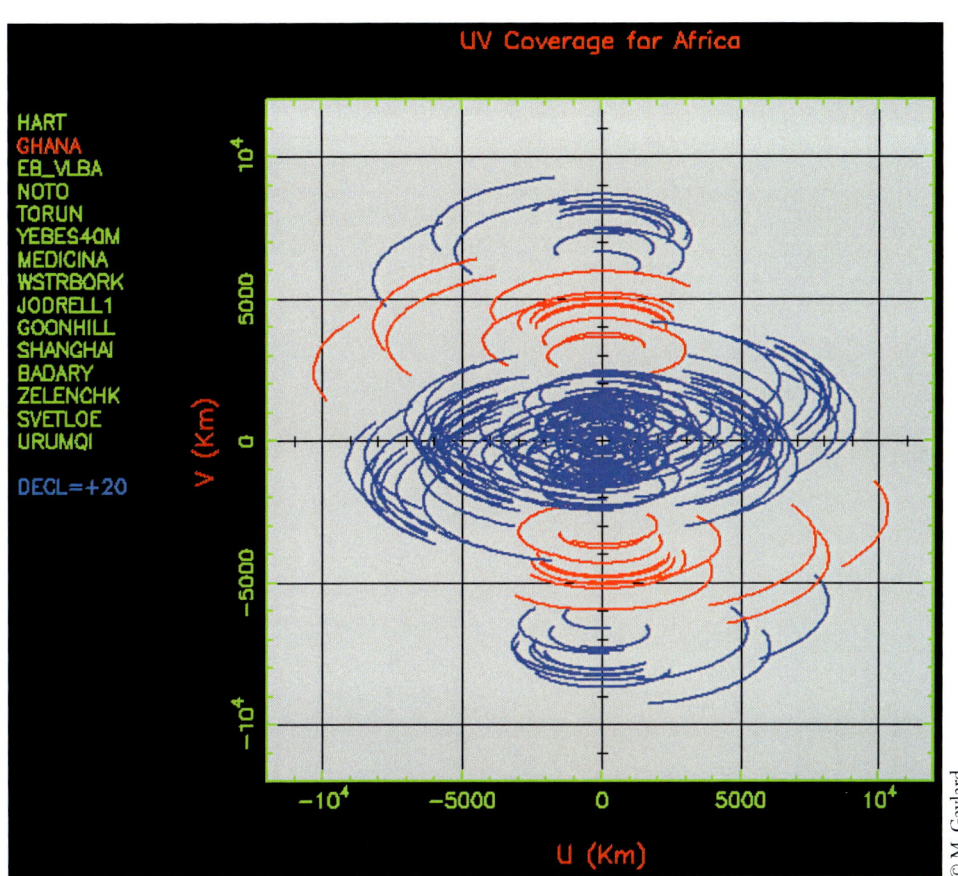

Figure 3: Adding Ghana, halfway between Europe and South Africa, fills in the big gap in coverage and will produce a far better radio image.

The KAT-7 at sunset.

world, and it was a South African idea.

The other telescopes – the XDM and the KAT-7 – have not been standing idle. While they do stand and appear idle from a distance, they have been doing some good science. The XDM is being converted into a permanent geodetic VLBI telescope, and the KAT-7 has been up and running since 2010, and is being commissioned – that laborious process of pushing the telescope to its scientific limits to check that it meets specifications. Within two years, the KAT-7 has already made some exciting discoveries, specifically observations of neutral hydrogen. The search for neutral hydrogen is one of the main reasons for building the MeerKAT and the SKA, as it will help us understand how galaxies are formed and what happened after the Big Bang (see Chapter 11: SKA science). However, for South Africa, this observation is important for reasons other than the science. It is the country thumbing its nose at Afro-pessimistic nay-sayers who say that Africa can't do real science. "The exciting results achieved by the KAT-7 have given us confidence that we know how to build a cutting-edge radio telescope in Africa to answer some of the fundamental questions in radio astronomy," Dr Fanaroff says. This is another example of his polite diplomacy. For the rest of us, the comment would be more along the lines of: "See what we did? What did you say about Africa not being capable of excellent science? Neh-neh-neh-neh-neh!"

☽

It is at this point in our story that various threads start to intertwine, weave together and form a large knot over a previously insignificant farm in the Northern Cape. The Losberg farm was identified by George Nicolson and Justin Jonas because it fits the specifications of the SKA – it had to have the right climate (clear skies, dry with very little rainfall); it had to be radio quiet with a small population in the surrounding area; it had to be relatively close to existing infrastructure such as roads, power stations, etc., because otherwise everything would have to be built from scratch; topographically it had to have hills to shield the site from external interference; it had to be accessible by roads. All these things had to be taken into consideration by the travelling SKA team.

Now the South African SKA Project Office is effectively split into three parts: one section was mandated to focus solely on the bid; the second was responsible for the MeerKAT; and the third had to drive the human capital programme (see Chapter 10: Benefits). Until the SKA Phase 1 comes online in 2019, the MeerKAT will be the largest telescope in the Southern Hemisphere. From the beginning, the MeerKAT was positioned as being independent of the SKA – even if South Africa lost the bid to host the SKA, it would still have the 64-dish array.

MeerKAT Project Process Manager Frank Curtolo draws the short straw and has to take a certain irritatingly inquisitive journalist to the Losberg site. Curtolo explains that the MeerKAT is something we have control over, unlike with the SKA bid, which was in the hands of Italy, the United Kingdom, China, the Netherlands and Canada. "The MeerKAT is funded by our government, we have the resources and skills, we have the site. That is what we must focus on, delivering MeerKAT as fast and as best as we can."

But South Africa's 64-dish telescope was also confusingly called a precursor because it would be on the SKA site, may form part of the first phase of the SKA if we got it, and its science case was strongly based on the SKA.

"The Department of Science and Technology and the National Research Foundation said that if we got to build a world-class facility, then we needed to do it for world-class science," Prof Jonas says. "And to do that, we had to look beyond the traditional science borders of South Africa, and so the SKA provided a nice, convenient way to interact with the international community. It had to have its own MeerKAT science case, so we looked at a subset of the SKA science case."

Although the new array's science case did include some areas of previous South African proficiency, a great deal of it was new. For example, the MeerKAT will look at pulsars, and since the early HartRAO days, South African radio astronomers have been investigating these celestial rotating lighthouses. But part of the science case has to do with neutral hydrogen, which was the impetus for the SKA – something South Africans had never looked at because other countries, such as Australia and Argentina, already had a huge head start in this field (see Chapter 7: African SKA).

But there is more to building a monstrous radio telescope than having a history of expertise and locating the perfect site. From the sounds of it, it is a logistical nightmare – which is why you need lots of people to get it off the ground. Yes, you have a site, but then you need roads connecting the site to the rest of the world. You need communications infrastructure, especially since you can't use cellphones. There is a sign, as you turn left off the provincial dirt road, requesting – in an authoritative tone that indicates it isn't a request so much as a pain-of-death instruction – to switch off your cellphone. You need power reticulation, workshops, offices, data cables both connecting the dishes and the site to the Cape Town office, linking roads on the farm, and – since it is more than an eight-hour drive from Carnarvon and then another hour and a half to the site itself – an airstrip to fly in all the people who need to accomplish the daunting to-do list above. "People", in turn, is shorthand for specialists in antenna structure, control and monitoring, computer software, computer hardware, and all the engineering disciplines you can think of from electrical to mechanical – which is difficult to say in one breath. These are the people tasked with anticipating problems that we wouldn't even think of. For example, swanky shiny electricity poles run from Cape Town to near the Losberg site. They carry electricity and fibre-optic bandwidth, and near the site itself the poles have to be made out of steel so that they don't interfere with the radio signals. The grandiose concept of a radio astronomy reserve as well as tiny details like that are the sorts of things that these specialists have to think about on a daily basis.

The offices, accommodation, workshops and correlator building are all shielded by Losberg, the hill – or mountain, if you're feeling very generous – after which the farm is named. Curtolo says: "It provides additional screening for the telescope itself; it is a natural barrier. That is how it was selected to place this facility in relation to the antennas." The hills surrounding the farm and the one separating the buildings from the antennas are the most effective tools for shielding the telescope from unwanted radiation. Additionally, the MeerKAT power and correlator building is sunken in order to minimise interference with the dishes.

Then there is also the antenna-manufacturing building. One of the main drivers for the MeerKAT (and the SKA bid) was technological innovation and skills development. We could quite easily have just put in a call to China and ordered thousands of dishes, but that would have defeated the purpose somewhat.[6] South Africa has the expertise to build these dishes and wanted to develop these skills, but also it wanted to keep the money in the country. The main buzz words in the country's political vocabulary at the moment are "job creation", and the construction of a large telescope goes a small way towards meeting that demand – the SKA SA project wasn't going to hand that sort of opportunity to another country. The KAT-7 has moulded dishes, and scaling up production with an existing mould is a relatively simple enterprise.[7]

But, as is the case when you're developing scientific

---

6 More than one person has noted that this was Australia's plan for building the dishes.

7 Well, this will be the case if the moulded-dishes concept is chosen. MeerKAT is being built through an open tender process, and companies from various countries are submitting bids.

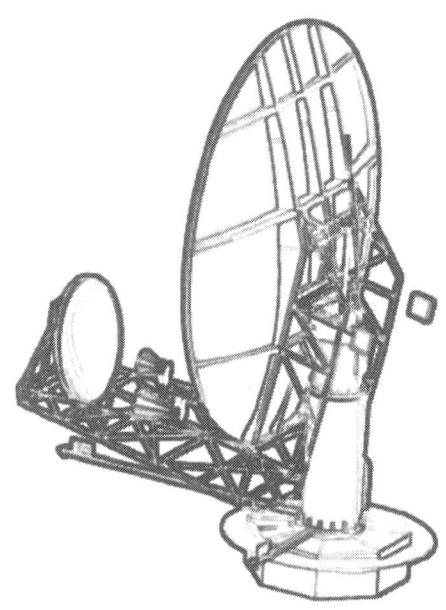

A design of a MeerKAT dish

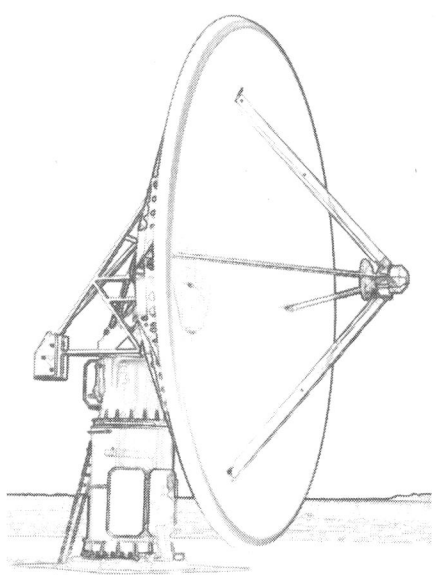

The KAT-7 dish

equipment, the MeerKAT antennas have evolved beyond the XDM and even KAT-7 dishes. The 64 MeerKAT dishes will be smaller than the XDM but larger than those of the KAT-7, at 13.5 metres. Also, it will be an "off-set" dish, which means it won't have a pointy nose like earlier incarnations when the secondary reflector is suspended over the middle of the dish, but rather a very strong jaw that will reduce interference with incoming signals.

You have to travel a distance away from the buildings to see the KAT-7 site, where there are only dishes, a container with *SKA South Africa* emblazoned on the side, and sheep. In the sweltering Karoo heat, the sheep seem particularly appreciative of the dishes, with more than 20 shoving each other to find a spot in one dish's shade. The sheep, for one, will be happier with the world when the MeerKAT is complete in 2016, mainly because the dishes will be slightly bigger, rather than substantially greater in number. Although there are seven dishes offering shade to a sheep that, given the heat, could spontaneously combust at any moment, they choose to fight over one.

The MeerKAT was designed to be a self-standing instrument, and is separate from the SKA, although it will eventually be incorporated into the mammoth interferometer. Once it had been established, SKA SA took large groups to view the site and Prof Jonas recounts what happened when he first took people there. One of the first guests was Steve Rawlings, a close friend of Prof Jonas and professor in Oxford's Astronomy Department, who tragically died early in 2012. Prof Jonas says: "Steve came on one of the trips, and he said, 'Build it and they will come. If you start doing this sort of thing, other people will show interest.' And he was right."

Aside from the KAT-7 and MeerKAT projects, the Losberg farm is attracting international interest as a radio astronomy destination. "Other people in the world started looking at [the site] and saying, 'You've got a site; we'd like to put down an experiment there,'" he says. PAPER and C-Bass experiments are already on the site. PAPER – otherwise known as the Precision Array for Probing the Epoch of Reionisation[8] – is an American project that will comprise 128 antennas. The C-Bass experiment – or, for a sibilant tongue-twister, the C-Band All Sky Survey South – has a sister telescope at the Owens Valley Radio Observatory in the United States, and is already in place at Klerefontein. The receiver for the telescope is being developed by a former Rhodes University undergraduate, now Oxford University doctoral student, Charles Copley.

So, for such a quiet site, there is a great deal going on – and this is even before construction for the MeerKAT has begun, let alone for the SKA.

---

8 Part of the SKA's science case, described in detail in Chapter 11: SKA science, includes investigations into the Epoch of Reionisation.

7
African SKA

The Square Kilometre Array isn't just a project, it's a mega-project, and that's not a catchy sensationalist title to attract people's attention. Governments, their departments, scientific and academic institutions, all of them have projects, some of them are even dubbed "flagship" projects. But as someone once said: "Not all flagship projects are born equal" – it takes more than money and a plan to have a mega-project.

There are a number of markers that help you identify a mega-project amongst its less impressive project cousins. It can leverage once-off government funding, attract investment and alter the identity of a nation. One example is the FIFA World Cup, hosted by South Africa in 2010. The funding wasn't taken out of the public works or the sports budget – it had a piggy bank of its own because it was recognised as an entirely different species of event. Big brands and industry also drank the water and lobbied against each other to sponsor the World Cup. And, perhaps most importantly, it altered perceptions of the country both at home and abroad. South Africa was now a country that was capable of playing in the big leagues, and had successfully hosted one of the biggest sporting events in the world.

Pick any government department and they will have a buffet of flagship projects, but what makes a flagship project a megaproject? Well, it's mainly money, scale and high profile. South Africa has had other sporting events, but the Soccer World Cup is in a league of its own. Why? Because of its sheer scale, bolstered by its profile and the amount of money that went into it.

Unfortunately, while it isn't difficult to get a country of soccer enthusiasts to buy into the excitement of hosting the biggest event in town, it's somewhat more difficult to capture the public consciousness when you're talking about radio astronomy. Their eyes glaze over because, unlike the United Kingdom or America or China, South Africa does not have a culture of science, where science habitually makes front-page headlines in newspapers and science *on dits* make their way into water-cooler conversation. But the World Cup and the SKA are the same beast.

Dr Fanaroff says: "Whereas you might struggle very hard to obtain incremental funding for astronomy or radio astronomy, mega-projects have a way of obtaining funding from different buckets. They don't come from the same place as normal funding." There is no way that the South African SKA bid could have been successful if all of the infrastructure, wages, advertising spend – and a number of other dependent areas – had to come out of the National Research Foundation's astronomy budget.

But it appears to be a bit of a chicken-and-egg question: which came first – the funding or the project? A project gains momentum as a result of the funding thrown at it, while funding can be leveraged as a result of its status. In the case of the SKA, it was a top-down decision. All of the SKA SA players say that there was always political buy-in,[1] that it was immediately recognised that this project should be a priority.

But the key to mega-projects is that they raise their profile and funding disproportionately. The estimated R2 billion allocated to the SKA bid is not proportional to the returns that it is bringing on a tangible level, ignoring the perceptions and how it repositions the country internationally and locally. From seeding astronomy as a nationwide science and the boost in the number of science, engineering and technology graduates to the trickle-down effect of infrastructure spend, the benefits are non-linear. Also, since the SKA SA project began and the country started sending graduates for training at overseas institutions, South Africa has become a leader in a number of highly skilled fields. For example, the country is now at the forefront of digital hardware, with the ROACH board – developed by South Africans in conjunction with the University of California Berkeley in the United States (see Chapter 9: The African VBLI Network) – used in radio telescopes all over the world; there is also antenna design, with South Africans recognised as leaders in this field. Those sorts of achievements are not proportional to the investment in the SKA project, and have a lasting impact.

Then there is also investment. Mega-projects are fertile grounds for nurturing private sector investment because of the profile, because of the exposure, and in this case because of the research and development. Global giants, such as IBM and Nokia, are getting in on the SKA action and working with the project office in developing some of the technology

---

[1] A number of Cabinet ministers throughout the project's history have given ready support. Naturally, the science ministers Ben Ngubane, Mosibudi Mangena and Naledi Pandor, and their deputies, have given the project their all, but other names keep cropping up, namely former finance minister Trevor Manuel, present finance minister Pravin Gordhan, former tourism minister Mohammed Valli Moosa, and former public enterprises minister Alec Erwin. These ministers are said to have supported the project, and recognised its importance. In something like this, it's quite important to have the finance ministers behind you.

required. From the perspective of the project, it benefits from the experience of world leaders and conversely it enables these companies to develop technology that doesn't exist yet and will inform their future products.

"It's not simply the fact that we are giving that much more money so we get that much more development. If you look at the Apollo project [in the United States], or similar high-profile projects, they spin off other kinds of development that wouldn't be spun off by incremental funding, no matter how much you increased the incremental funding," says Dr Fanaroff.

For Dr Khotso Mokhele, who oversaw the inception and building of SALT and seeded the SKA project in South Africa, the importance of a mega-project is about recasting the country's identity, and repositioning it as being on equal-footing with developed countries. He talks about people who live in cities that didn't have stadiums, who didn't get jobs, who were living in a rural town in the North West and didn't get to watch a game – they still took pride in being part of a nation that was able to bring the world to come together to play soccer in their country. "Those kinds of investments [for the SKA and the World Cup] become important not only because they put bread on somebody's table today – because many of them put very little bread on the national table when the demands for bread are as huge as in this country – but they are projects that make a nation look at itself and say, 'You know, we have myriad difficulties, challenges and problems, we are on the southern tip of a continent with all these issues ... but we belong.'"

He argues that, while it may just be a telescope, to South Africa it is more because it symbolises hope. "Again, no bread on the table for the poor people in this country. Kids who study under trees don't all of a sudden get books and find themselves in classrooms because we are building the big telescope," he says, emphasising that this doesn't make these challenges less urgent. "But, my goodness, they may be studying under a tree, they may be sharing a classroom with different grades, but if we expose them as young kids to the value of having this instrument in the country, even under those difficult circumstances, their perception of what they are capable of must change."

Without these mega-projects, South Africa sentences itself to the fate the White Paper on Science and Technology warns against: that we will become "a second-class nation, chained forever to the treadmill of feeding and clothing ourselves". As MeerKAT Project Manager Frank Curtolo says: "Without hi-tech projects, then you resign yourself to being a Third-World country. The moment you say, 'I want to strive to become First-World,' you need hi-tech projects. They drive innovation and development."

Something that South Africa's SKA bid threw into the spotlight is how some developed nations perceive the country. The Australian Science and Technology Minister Chris Evans said in the run-up to the bid announcement that the main impediment to the Australian bid was the aid-mindset in Europe, that South Africa may win because it was a way to help Africa. Science and Technology Minister Naledi Pandor's restrained response – although you wonder what she really wanted to say – was: "The most recent comments attributed to Senator Evans reflect a very inadequate understanding of where Africa is today. Various economic analysts have confirmed that Africa is a vibrant economic growth region … Our bid is sound and we won't insult any party in an effort to sway decision-makers."

And that isn't just political polemic, but it would appear that overseas this data has been obfuscated by an image of Africa with a begging bowl. Mega-projects are one of the few ways to alter that perception. According to Dr Fanaroff, the SKA project has a dual message: that Africa requires the skills and capacity to become a major economic growth story, but also that the continent is seen as the next great business destination. "You have to have people who can absorb the technology that the global economy wants to sell to Africa," says Dr Faranoff.

☾

There is a reason that South Africa's SKA team spent weeks traipsing around in the middle of nowhere. The site requirements are quite specific: you need a pristine radio astronomy environment with no people that can house a couple of thousand radio dishes and other antennas. To clarify, it is not possible to have all 3000 dishes – which is only one part of the SKA – on the Losberg farm. The farm's big, but it's not that big. In fact, the Northern Cape isn't that big.

It is difficult to say what the SKA will look like, because – well – it hasn't been fully decided yet. There is now the additional complication that the radio telescope will be split between South Africa and Australia. The international SKA body is in the process of deciding what guise it will take. The form of the SKA will depend on which technologies are chosen by this international consortium.

To give you a vague idea of one possible SKA – the SKA Organisation has yet to give the definitive word on what it will look like – its heart will be at the Losberg farm, but it will sprawl up through the rest of Africa, extending 3000 kilometres away from the core site. About half of the telescope's instruments will be in the core, with the rest forming what looks like a star on the African continent radiating out from the Northern Cape.

In news media, the main focus has been the dishes, but there is a lot more to the SKA than the dishes. The radio wave part of the electromagnetic spectrum is unwieldy, ranging from tens of Megahertz to hundreds of Gigahertz,[2] which means you need instruments focusing on specific parts of a rather large spectrum. So the SKA – which will observe from 70 MHz to 10 GHz –

---

2 Some context: 10 MHz to 100 GHz is the frequency band. In terms of wavelengths, this is 30 metres (10 MHz) to 3 millimetres (100 GHz), which is like comparing something that could fit between the teeth of your comb to something longer than the width of an Olympic-size swimming pool.

is actually made up of three different cores: a low-frequency array and two mid-frequency arrays. The one everyone knows about is the dish array, mainly because it's easier to explain radio astronomy in terms of a large receiving dish, and because – well – they look cooler. Radio dishes, as seen in movies, are the instruments that people discover aliens with, that secret agents have fights to the death in, and just look more impressive. There is also the fact that these dishes will spiral outwards from the SKA core site and so there are also more of them.

South Africa will host this part of the SKA, which means that its eight partner countries – Botswana, Namibia, Ghana, Mozambique, Mauritius, Madagascar, Kenya and Zambia – will also have satellite stations.

There are other arrays, although the exact form they will take hasn't been decided. The low-frequency array is an aperture array that looks a little like wilting flowers or helicopters whose blades haven't started rotating yet – a central pole with four blades akimbo. With the traditional radio dish, the signals come in from space, bounce off the parabolic dish and are collected in the telescope's nose; with aperture arrays, the signal is received as soon as it touches the antenna. The mid-frequency aperture array looks very different from the low-frequency aperture array – the mid-frequency array looks like a hi-tech collection of lily pads on a sandy Karoo pond, jagged white circles on a beige background. "Circles" might be a bit generous, actually: they look more like you've told someone to draw a circle using only straight lines.

Each of these arrays has a "core", which is five kilometres in diameter, and will sit side by side, not quite touching, in the Karoo heat. This is why the government bought the Losberg farm, with its impressive 14 000 hectares – how else could you fit these instruments into one space? But in order to increase the SKA's coverage of the sky, the dish array spirals beyond Losberg's borders in five arms that extend as far as 3000 kilometres.

There will be satellite stations – each comprising about 30 antennas – in South Africa's eight partner countries, although the number and where they will be located in these countries are points still under discussion. As things stand, there should be three in Namibia, four in Botswana and one each in Mozambique, Mauritius, Madagascar, Kenya, Ghana and Zambia. The governments of these countries, as well as the African Union as a whole, have thrown their weight behind the project, with the African Union officially calling for support for the SKA.

The support is not surprising, because having part of the SKA in your country has ancillary benefits (see Chapter 9: The African VLBI Network). Aside from the obvious prestige, human capital development and foreign direct investment, there is the matter of connectivity. The SKA may comprise thousands of antennas, but it operates as one instrument: for a disturbing futuristic example, you can imagine it as an army of robots moving as one entity. The reason this is important is that all the data that these individual antennas collect has to be added to the data collected from all the others, which will happen at a correlator at the core SKA site.[3] Behind and underneath the antennas, an invisible tapestry of fibre-optic cables will transport the staggering amount of data back to SKA HQ. This radio telescope will accrue more raw data in one week than humankind has generated in its entire history. The benefit for the partner countries is that the infrastructure required for this data connectivity has trickle-down effects, and will also boost the Internet capacity in that country.

The SKA will now include infrastructure in Australia and South Africa, including the precursor telescopes ASKAP and the MeerKAT. Australia gets the low-frequency array, while South Africa gets the mid-frequency aperture arrays. However,

---

3 A correlator is a supercomputer that adds all the signals together. At the moment, the technology doesn't exist to correlate all the signals. See Chapter 12: Looking to the future.

the mid-frequency section is bigger, so the split was more like 70/30 to South Africa than an equal split. While many South Africans were ecstatic at the news that South Africa would be splitting the SKA with Australia, others chose to see it as an unfair compromise and "a kick in the teeth for Africa on Africa Day".[4] But the reality is that a split was always the most likely outcome. It was just a question of which country got what percentage. Being awarded 70% of the SKA is, observes Dr Fanaroff, "a huge victory" for Africa. "It recognises the fact that Africa was recommended [by the independent SKA site advisory committee] and that 70% of the SKA will be built in Africa. It shows great confidence in Africa," he says.

The SKA will be bigger than the Large Hadron Collider (LHC) at the European Organisation for Nuclear Research (CERN), which is the largest scientific instrument built to date, as well as one of the most expensive.[5] The LHC is a particle accelerator, whose 27-kilometre circumference straddles France and Switzerland, but the SKA will cover an entire continent. Interestingly, there are strong parallels between the SKA and the LHC: they are both trying to answer similar questions. How did the Big Bang happen? What is dark matter? What are black holes? While the SKA will look at the bigger picture, trying to understand how the universe works from the movements of the galaxies, the LHC tries to divine the same answers from the fine print, written in the way particles interact.

The LHC was built by CERN to test previously theoretical predictions about the behaviour of high-energy particles. It is a collaboration of more than 10 000 scientists from more than 100 countries. The collider accelerates protons or lead nuclei around the tunnel and smashes them into each other. By doing

---

4 25 May, when the decision was announced, is celebrated as Africa Day.
5 The International Space Station, at a cost of about $150 billion, is technically the most expensive man-made scientific experiment, but that's in space so it doesn't count. The LHC has a price tag of about $9 billion.

this, scientists hope to discover the smallest particles in the universe, namely the Higgs Boson particle, otherwise known as the "God particle".[6]

The SKA will not be as controversial as the LHC. Since the LHC involves fiddling with things we don't understand, people were terrified something would go wrong and, well, annihilate the world. Some folk in the United States were so worried about possible mishaps that they've been in and out of court trying to halt operations at the LHC. In 2010, their second appeal was thrown out. The judge's judgment is priceless. His main reason for dismissing the group's appeal was that America doesn't have jurisdiction over CERN's operations. However, he didn't stop there: "Accordingly, the alleged injury, destruction of the Earth, is in no way attributable to the US government's failure to draft an environmental impact statement." This is the equivalent of saying: "If the world does implode as a result of this machine, we would like it on record that it wasn't us!"

---

6 The Higgs Boson particle appears to be more elusive than Big Foot and the Loch Ness Monster combined – and if we'd spent $9 billion searching for them, we'd have found them by now. The Higgs Boson particle is allegedly what gives all matter its mass. According to physicist Peter Higgs, there is a field – called the Higgs field – that permeates the space between every particle in the universe: imagine iron filings in a magnetic field, but on a super-massive, universal scale. But different particles interact differently with the field. An electron, for example, having little interaction with the field, moves very quickly and so has a small mass. A particle that trudges its way through the field moves slowly and thus has a larger mass. In physics, a field is associated with a particle, such as a photon with an electromagnetic field. So to prove the existence of the Higgs field, scientists are searching for the Higgs Boson particle. How do you do this?

You accelerate two streams of protons around a tunnel at the speed of light and smash them into each other. And then scientists trawl through the wreckage (which only lasts for fractions of a second) and search for traces of exotic, previously undiscovered particles that were once there … well, may have been there.

A prayer to the young Moon

|Kaggen made the Moon, and like everything that |Kaggen created, the Moon is wise and crafty, and the animals listen to him. If you do not ask him for help before you go hunting, he will make sure you return home with sore feet and empty hands. Your arrows may be sharp, and your footfalls as quiet as a feather, but if you don't ask the Moon for help, you will kill nothing. So, before we go hunting, we implore the Moon to whisper a word, so that we can eat. This is what we sing:

Young Moon!
Hail, Young Moon!
Hail, hail.

Young Moon!
Young Moon! Speak to me!
Hail, hail.

Young Moon!
Tell me of something.
Hail, hail!

When the sun rises,
You must speak to me,
That I may eat something.

You must speak to me about a little thing,
That I may eat.
Hail, hail.

8
Challenges

You'd expect Afro-pessimism from Australia – South Africa ended up being the dark horse in the race, a lightweight that surprised the world. Australia is a world leader in radio astronomy and never thought it would be threatened by a relatively small country on the tip of Africa. But it is very disheartening when you realise the extent to which that sentiment simmers within South Africa.

Initially, you only heard rumours of international Afro-pessimism, with the people involved telling you that this negative perception stalks diplomatic and scientific corridors. But this sentiment burst like a lanced wound in the run-up to the site announcement. Perhaps it is naïve, but you really hope politicians would know better than to let their personal sentiments spill over into the public space, and that they would be canny enough to watch their tongues in front of journalists. At the beginning of 2012, things were already starting to look bad for Australia.

It all starts with the Square Kilometre Array site advisory committee, the independent technical committee mandated to adjudicate the suitability of the two sites. In February 2012, this committee recommended one of the sites to the SKA Founding Board; while it was just a recommendation, it was important – no matter how much it was later downplayed – because this was an independent committee that made a recommendation on technical grounds, and it could indicate which country was going to win the bid . It is easy, in retrospect, to say, "Well, it was so obvious", but in the build-up to the announcement anything seemed possible, especially since no one would talk about what the recommendation said. A few weeks after the clandestine report was handed over to a select few, none of whom were journalists, two Australian newspapers published that South Africa had been recommended over Australia. Most journalists covering the SKA were on the phone, harassing the national and international SKA officials, but it was as successful as trying to lick your elbow. At the time, Justin Jonas, associate director for science and engineering at SKA SA, said: "The report is still confidential and, until we know otherwise, we are not going to say anything." We were all back to square one: no one was talking, although now there were the unconfirmed newspaper reports.

But then the Australian Science and Technology Minister Chris Evans stood up in front of the Australian Press Club and said that the main impediment to the Australian bid was an "aid mindset" in Europe. "The thing that works against us the most is the sympathy for doing more in Africa – the European view that says we ought to be doing more development in Africa," *The Australian* quoted the minister as saying.

People in South Africa were outraged. Although he later said that he had been misquoted, Senator Evans' comments typified the afro-pessimism that people had said existed, but which was brushed off as over-sensitivity.

The underlying message here is that it was possible Australia would lose to South Africa not because the African country was technically superior or capable of excellent science, but because of sympathy. These comments reinforce a stereotype of South Africa as a country with a ready begging bowl. But perhaps more concerning than remarks from a minister who was pre-emptively justifying the loss of his country's SKA bid were the comments that came out of South Africa. On most South African news websites, the comments section is often longer than the story.[1] This is an example of the comments in response to an article about the senator's Afro-pessimism: "Just maybe Mr Evans are [sic] saying things that should have been said a very, very long time ago?? Political correctness is fooling oneself!!"

This internal negative perception of the country's ability to host the SKA says a great deal about our national psyche. It is not an isolated comment. Here are some others:

"Wonder how many astronomers will get taken out in the remote area they plan to operate in – have sent them a complete file of the farm murders, rapes, robberies and muggings – advising then they will be much safer on Oz. Not even a lepton brain would want to invest in such a dangerous place as SA."

"Don't hold your breath. The choice is between an organised sophisticated and stable economy in OZ or a country whose administration in government is in disarray with corruption, graft, fraud and lawlessness. Which would you choose? One fact I know is that when SA lose to Aussie (again!) – the racists are all going to be shouting that it was a racist anti-Africa decision. Any bets?"

---

1 Often these comment threads become very heated, and people feel the need to write hundreds of words in response to their nemesis of the moment. The author wonders if they feel the irony of moaning about a lack of productivity in the country while spending hours entering into pointless arguments online.

"And how many houses could have been built with the billions that will be spent on trying to hear the voice from outer space? How will that voice enhance alleviation of poverty, poor infrastructure, failing healthcare system and education? Let us get our ducks in a row, sort out the problems that plague us and once we have achieved 100% literacy and world-class public healthcare facilities then we can play with big toys to try and hear that one voice from outer space."

In 2011, Deputy Science and Technology Minister Derek Hanekom said "Afro-pessimism is real … We believe we are the most suitable country to successfully host the SKA, we compete with the best in the world. But we are proceeding anyway and great astronomy is going to be done with or without the SKA."

In the context of the above comments, it makes you wonder whether Minister Hanekom was talking about external Afro-pessimism or the canker in our own country. One armchair commentator said that if South Africa lost the bid, then it would be hailed as racism, but never considered that if Australia lost it would be seen as preferential treatment for Africa.

But what the nay-sayers fail to realise is that if South Africa wants to be an international player, and lift as many of its people as possible out of poverty, then it needs to play in international games. Khotso Mokhele likens the scenario to playing tennis: "People who play tennis don't particularly like playing with people who don't play as well as they do. We all like to play with somebody better because if you are good, you don't want to play somebody who hits the ball over the fence every time, because then you have to go and fetch the ball on the other side of the fence."

This is what South African science has to appreciate, he says. "If you want to partner with the best and the brightest in the world, they have to believe that by partnering with you they get something from you. They are not going to fetch the ball on the other side of the fence for you all the time. So, yes, we have a small science system, there have been many punches above its weight, but we have to have those pockets of excellence because the best only want to play with the best."

If South Africa wants to play a role as a leading global nation, then it has to interact with its peers as an equal – and in science, this means investing in science.

South Africa needs to realise that it is capable of hosting something as prestigious as the SKA. There is something deeply concerning about a country that refuses to believe that it can achieve great things when it has the means. Fortunately, these sentiments are not shared by South African decision-makers. "The Department of Science and Technology has been pushing the idea of a knowledge economy," SKA SA director Bernie Fanaroff says. "The business community likes the vision of science and technology as a vision for the future of South Africa. In our interactions internationally, the idea of science and technology being an important issue for development has been well received across the political spectrum."

The White Paper on Science and Technology highlights a knowledge economy as a way to drive economic growth, skills development and competitiveness: "Science and technology are considered to be central to creating wealth and improving the quality of life in contemporary society." Societies advance through scientific endeavour, research and development. If this country were to decide that its only priorities should be basic needs, then it assumes "the view that we are a second-class nation, chained forever to the treadmill of feeding and clothing ourselves".[2]

The SKA will not solve the identity crisis that many South

---

[2] This is taken from the White Paper on Science and Technology, and cannot be repeated frequently enough because this will be the South African reality if the country ignores scientific research. Without technological advancement, a country will always lag behind other, more developed nations. Published in 1996, this far-seeing paper envisioned where the new democratic South Africa – despite the entrenched inequalities of its apartheid past – should be in the future: a country on equal footing with the rest of the world.

Africans seem to be suffering, but it creates hope that they will be able to break out of this mould of Afro-pessimism. There are very real problems in South Africa, from political wrangling, corruption, crime and others besides. But sitting above these problems is also the fact that the country has niches of excellence, with talented people doing world-class science, as well as political will to play in the international scientific arena.[3] Yes, there are problems, but South Africans also need to realise that their country is capable of great things, such as hosting the largest scientific instrument on Earth.

When asked whether the SKA would be "good" for South Africa, Dr Fanaroff asks: "The problem of course is what do you mean by good for our people? Some people interpret that as it will create jobs tomorrow, build houses tomorrow, improve health tomorrow." While the SKA might do some of that, "its more important impact is the much longer-term impact of changing both our own perception of what's possible and the world's perception of what Africa can do, and what South Africa in particular can do. It will also create a significant critical mass of young people and industries with the kind of capabilities that will enable them and South Africa to play not just a role, but a leading role in global technology," he says.

☽

It cannot be denied that South Africa has problems. Afro-pessimism is an inability to see anything beyond those problems, and in the South African context, it is a refusal to acknowledge that the country is relatively schizophrenic: a world leader in niche areas, while struggling with widespread poverty and inequality. In mining, medical isotopes and coal-based synthesis and gas-to-liquids technologies, South Africa is a world leader, but at the

---

3 Some South African trivia: The CAT scan was developed in South Africa, where the first heart and dual heart-lung transplants were performed; and the Kreepy Krauly swimming pool cleaner is a South African invention.

same time the words "dismal" and "crisis" are synonymous with its education system.

Once again, one radio telescope – even an incredibly big one – cannot solve a country's problems. It would be glib and insensitive to try to condense the problems in the South African education system in a couple of paragraphs. It is a complex and difficult situation with no easy resolution. But when the country decided to put in a bid for the SKA, it needed to develop the technical skills that would make it a real contender. In the stiff competition to host the billion-dollar instrument, it would – and rightly so – be the first thing competitors highlighted to deride South Africa's bid.

Justin Jonas, the man who initially carried the country's SKA torch, says that once South Africa had decided to submit a bid to host the telescope and had begun thinking about an instrument of our own, "at the same time, some of us were thinking, well, we need to build capacity. We don't have many radio astronomers; we're going to need more, especially on the engineering side. We're going to need engineers who understand radio telescopes." The country had engineers, but engineering is a very diverse field and very few of them knew the intricacies of building radio telescopes.

Serendipitously, at about the same time, the National Astrophysics and Space Science Programme (NASSP) was getting off the ground. "We were looking at what South Africa needed when SALT came online. What could we do to make sure that we were really ready for SALT and the next project after it, which turned out to be the SKA?" says Patricia Whitelock, the acting head of the South African Astronomical Observatory who is also the acting director of NASSP. Dr Whitelock is a small lady with a warm smile. "We decided we needed to train astronomers in South Africa. At the time, there were very few astronomers being trained in the country ... there was one at the University of Cape Town, occasionally one came out of Rhodes, occasionally one from Wits or Potchefstroom; but they hardly spoke to each other."

Astronomy, even globally, is a very small discipline, which makes connections vital. While the love of a particular aspect of astronomy may dictate someone's career, you are usually pushed into a discipline because of the opportunities at the time, and the people who you know. Dr Whitelock says that the reason she went into optical astronomy rather than radio astronomy or another area was because of the opportunities available at the time. Her work now is still affected by the people with whom she studied. "I was trained at London University," she says, "and I still keep in contact, and often collaborate, with people who were undergraduate and postgraduate contemporaries of mine. They are now my collaborators ... some are in the United States, some are in the United Kingdom; the previous director here [the late Bob Stobie] was an undergraduate contemporary of mine. I knew him since we were both 18."

This is why collaboration and the idea of an African astronomy "team" are so important. Hosted at the University of Cape Town, NASSP was a way of "getting African astronomers together to collaborate and support each other, and get the people together who were going to do radio or optical [astronomy] or theoretical work, so that they were a team right from the beginning," says Dr Whitelock.

NASSP took in its first year of students in 2003, which was coincidentally the year Bernie Fanaroff was brought on as director of the SKA SA Project Office and South Africa's bid became "serious". The astrophysics programme was meant to develop skills in astronomy and create a sense of unity among students, but the SKA was big – and it needed a human capital development programme that would focus on SKA science and engineers. "We wanted something that was more specific for us, for radio astronomy," says Prof Jonas. "That grew into what has become our human capital development programme. It started off very small, with literally just a few bursaries, but now it's become a big programme where we go all the way through to research chairs."

Research chairs are a relatively new idea in South Africa, having been implemented in 2005 after being copied from countries such as Canada. The idea behind them is to have leading academics at South African tertiary institutions whose primary activity it is to do research and train postgraduate students. By 2012, the Department of Science and Technology – through the National Research Foundation – had invested R1.1 billion in the initiative, with a number focusing on astronomy and SKA and MeerKAT science.

This feeds into both the NASSP and the SKA human capital programme. While unifying students, NASSP aims to fill in the gaps in their knowledge by exposing them to other areas of astronomy. "There is the sociological side of getting people together, but it was also about getting the best teachers available in South Africa to teach this programme," says Dr Whitelock.

With the research chairs initiative, the best teachers available now also include international leaders in their fields, many of which were not previously taught in South Africa. As Prof Jonas says, "The human capital programme will focus on the science areas of MeerKAT, so, for example, we started attracting world-renowned people who had neutral hydrogen experience.[4] It's a very common science area internationally, but we didn't have it here locally."

Between 2005 and 2012, the government has spent R55 million on bursaries and scholarships for the SKA SA's human capital development programme, with 22 doctoral and 66 master's graduates to show for its cash already. By 2012, it had awarded 398 SKA SA postdoctoral fellowships and PhD, MSc and undergraduate bursaries, 70 of which went to Africans outside South Africa. This may not seem like a lot in the global context, but they are staggering figures for the South African astronomy community. While the South African academic community is relatively small, its pre-2005 astronomy

---

[4] By studying neutral hydrogen, astronomers may be able to discover what happened just after the Big Bang. See Chapter 11: SKA science.

community was minuscule. Partly because of its success and partly because of the high profile of the SKA, another R200 million has been allocated for 2012–17.

This investment has also spilled over South Africa's borders, because other African countries are included in the human capital programme. The African SKA includes Botswana, Ghana, Kenya, Madagascar, Mauritius, Mozambique, Namibia and Zambia. Madagascar and Mauritius already offer postgraduate courses in astronomy and three of the other partner countries have undergraduate programmes. In the very near future, astronomy will be taught at institutions in all eight partner countries at an undergraduate level.

But the human capital project isn't just about astronomy or astrophysics. The SKA project needs many other skills, such as engineers, software developers, technicians, and many more besides.

Because of the education situation in South Africa, if you want highly skilled people, you have to train them yourself. And this problem isn't confined to astronomy; it is endemic to most highly skilled areas in South Africa. The head of a South African science agency once told me that "the skills just aren't there … We advertise positions and we can't find people. We have a high unemployment rate, but also poor skills."

HartRAO MD Mike Gaylard discusses the skills shortage with respect to the observatory in Hartbeeshoek: "There is a shortage of skilled people. Trying to recruit skilled people, really hi-tech people who can do the things that we need, you basically have to build them yourself."

So while a number of SKA-funded students have been fed through NASSP, many of the bursaries have gone towards engineering. But even then, radio telescope engineering is an arcane and specialised field, and the world's best brains in the discipline are overseas. "We set up our MeerKAT office, to build MeerKAT, with a lot of young, but inexperienced radio astronomy engineers," says Prof Jonas, adding that they needed to be brought up to speed on what was happening internationally. "So we placed people in different institutions around the world, so we could get them going."

Dropping fresh young South African engineers into the boiling water of international institutions has yielded incredible results, the most notable being a collaboration with the University of California Berkeley in the US. "On the digital side, we identified Berkeley as a place where we would send people because it had a big digital group going," he says. "We decided to plug into that collaboration. And that's been a huge success. We've more than plugged in. We've taken over." This collaboration resulted in the creation of the ROACH board, which stands for the "reconfigurable open architecture computing hardware". The ROACH board will correlate all the MeerKAT's data together. This piece of hardware has attracted so much international interest that it is now used in many radio telescopes around the world, and it has South African fingerprints all over it. It is still to be decided, however, if this technology will be used in the SKA.

Those sorts of projects are why the SKA SA project is able to retain such incredible talent – because it is exciting. The SKA is a logistical and scientific challenge, with "challenge" being a euphemism: people on the outside might even go as far as to call it a nightmare. Nothing like this has been done before; it is sprawling, technical and will achieve things we've only imagined, using technologies that do not exist yet. You and I might say impossible and go to the pub to think about such a daunting project over a beer, but people involved in the project rub their hands together in anticipation, excited to break new ground.

"Young people need exciting programmes," Dr Fanaroff says, and that is what the SKA offers them. In 2012, Google was voted the best place to work at in *Fortune Magazine*, and in an interview in that magazine, Google co-founder and CEO Larry Page said: "If you're changing the world, you're working on important things. You're excited to get up in the morning. That's the main thing. You want to be working on meaningful, impactful projects, and that's the thing there is really a shortage of in the world."⁵

The same sentiment makes young people eager to work on the SKA: they are doing something challenging and meaningful. Prof Jonas agrees: "It's the case when you have a mission-led project. If you have a project where there is a very strong goal, people love working on those projects." From what he says, young people would choose goals and excitement over money. "Most people are working for 5% to 10% lower than what they can earn in industry, but they are prepared to because it's a hell of a lot more exciting than anything they'd be doing in industry. It's an international project."

But, at the same time, the SKA is not forever – once the project is built, the operational SKA will not have positions and projects for all of these engineers and technicians. Prof Jonas takes a rather pragmatic view. "Intellectual property walks on two legs," he says, hastening to add that the original quote is not his and that he wishes he could remember who has said it to him because he says it often. "It is not the specific technology that you are developing. It is the people who you are developing. So don't worry about patents and all those sorts of things – worry about the people who you develop."

It is seemingly incongruous, but the plan is not to retain the skills. This is not unique to the SKA. A number of South African science councils and companies see their role in the democratic dispensation as training grounds. Contrary to the private sector – which, if it could, would keep highly skilled people under lock and key – South African science councils and agencies are mandated to train up young graduates, and then allow them to leave. For example, the Council for Scientific and Industrial Research (CSIR) CEO Sibusiso Sibisi says that human capital development is one of the main thrusts of the institution. With

---

5 Adam Lashinsky, "Larry Page: Google should be like family", http://tech.fortune.cnn.com/2012/01/19/best-companies-google-larry-page/, 19 January 2012.

a staff complement of 2326, the CSIR is the largest research institution in Africa. "Some may stay at the CSIR, some may move on, joining other institutions setting up businesses ... We are consciously working to build a knowledge base in the country," says Dr Sibisi.

The same thinking infiltrates the SKA project. "At the moment, if you look at the retention in the system, we retain a big fraction of our students coming through the human capital programme into the project. That is because we still need people," Prof Jonas says. But in a couple of years, this demand will have dwindled. "The intention in the end is that we will probably only skim off 10% of them. About 90% of them will go off into the market."

When the SKA is finally built, many of the present SKA employees will have moved on. Many will go into industry, others will sign up for other exciting scientific projects, some may start their own businesses. It is difficult to gauge the impact of this human capital development programme on South Africa. Prof Jonas acknowledges this: "The track back is almost impossible. But you know that you are doing a good thing; you somehow know."

☾

It takes three attempts to get hold of Henk van Wyk, the head of Agri Noord-Kaap, a farmers' union in the area. First, it was a problem with cellular reception; then the Telkom line wasn't working. This is nothing out of the ordinary for the Northern Cape. "We have a problem with Telkom," he explains.

At the heart of the problem is the fact that the Northern Cape is very remote: vast distances and a sparse population mean that it is often uneconomical to provide telecommunciations services to the area. There is the usual wear and tear on the infrastructure, but the main problem is cable theft. Telkom isn't the only victim – Transnet, the state-owned freight transport company, also feels the bite of Asia's demand for copper cable, which feeds the black market in South Africa. It is estimated that the country loses about R20 million a year to copper theft, and Telkom has lost nearly R2 billion over the past four years due to copper theft,[6] ignoring the loss of reputation, especially amongst Northern Cape residents.

"Thousands of kilometres of copper wire were stolen and Telkom is not interested in fixing it, it's not economical," Van Wyk says. This means that most farmers rely on cellphones and radios to keep in contact, although it must be said that cellphone reception in the province is a bit like a mirage: sometimes there, sometimes not, and always insubstantial.

Now add a large radio telescope that requires radio-quiet conditions and no cellphone towers and you have a crucible of discontent, with Van Wyk saying that many farmers lost signal because of the radio telescope projects in the area. Whether or not South Africa got the opportunity to host the SKA, it would still have the MeerKAT, and the legislation protecting it is here to stay.[7]

The Northern Cape already has seemingly insurmountable obstacles (see Chapter 10: Benefits), and being unable to communicate with the rest of the country is only going to exacerbate the situation. There is also concern among farmers that no communications means they will be vulnerable to farm attacks,[8] as they will not be able to call for help if any-

---

6 The data was reported in *Business Day* on 16 March 2012.
7 The government passed the Astronomy Geographic Advantage Act of 2007 to preserve the radio quiet in the area.
8 While there is some uncertainty about the exact figures of farm attacks and murders, there is no doubt that it is a serious problem. National commercial farmers' organisation TAU SA estimated in 2012 that there have been about 1550 farm murders, although no specific timeframe had been given. AgriForum says there have been "at least 2617 separate farm attacks, in which some 1445 people were murdered", but that the number could be higher. The Northern Cape has very few incidents of this nature, but that does not mean that the fear of farm attacks is not a very real one for South Africa's white farmers.

thing does go wrong on their farms.

A 2009 report by the University of the Free State on the socio-economic conditions of the area says Northern Cape inhabitants are in an invidious position. "A choice between Telkom and its cellular 'rivals' is a choice between the devil and the deep blue sea. None of these organisations has the first notion of customer service and the contempt with which they treat their platteland client base is truly breathtaking," the report reads. At the moment, though, these are the only available options.

The report quotes one farmer, Willem Symington, as saying: "Currently, the telecoms services consist of partial cellphone coverage, outdated Telkom services (manual exchanges) and radio systems. We are battling to cope in this context … We would like to support the [SKA] project. We realise the huge value of the project for South Africa. However, this cannot happen at the cost of our social and economic survival."

This problem, however, pre-dates the MeerKAT and SKA, and SKA SA director Bernie Fanaroff is very straightforward when asked about the tensions between SKA SA and farmers in the Northern Cape. He says it all boils down to expectation management and what a radio astronomy project can and cannot deliver. "We can't deliver services," he says. "It's not only among the poor constituents, but also among the wealthy farmers. They haven't had adequate telecoms in the past because Telkom has failed to maintain the lines there 'cause they say it's not economical." Similarly South Africa's cellphone service providers have not extended their coverage because it's also not financially viable, he says.

"So some farmers say we'd taken away their services – which is generally not true[9] – and have tried to leverage us to deliver services they would otherwise not have access to. We help where we can, but we can't solve the area's problems."

Telkom took great umbrage at the allegation that the company

---

9 Although not all farmers near the MeerKAT and SKA site were affected by the preparations, about five to 10 were.

was ignoring its Northern Cape clientèle. "Cable theft is an ongoing problem ... a major inhibitor to Telkom's capability to maintain and improve service levels," says the company's spokesperson Pynee Chetty. "The recurring nature and extent of these thefts have resulted in certain areas being identified as ... 'hotspots', making it no longer commercially responsible to continue replacing stolen cables." He says that Telkom is developing a satellite to provide for these customers with their unique needs — no copper cable and radio quiet. "Customers who are currently serviced by manual exchanges, old copper-wire line technologies and customers who fall within the SKA exclusion zones will be serviced by this initiative."

However, the only problem with Telkom's strategy, which does appear to be the best solution to a rather difficult situation, is that none of the Northern Cape farmers knows about it. Van Wyk, who is in contact with most of the farmers in the area, says he has not heard of Telkom's plans. But satellite technology does seem to be the way to go. According to Van Wyk, satellite company Grintek would be conducting trials of its satellite technology in June (2012). Price will be the "deciding factor" in choosing companies' technology. "But [Telkom's] handling of the matter wasn't fair. We still haven't had answers from Telkom," says Van Wyk.

The SKA SA Project Office has not been idle either in trying to find solutions to the situation. Dr Adrian Tiplady, the South African SKA site bid manager, says that the project has developed a number of alternative communications for the local community. "We have been piloting 'radio astronomy friendly' technologies – telecommunication services that operate outside of the operating frequency range of the SKA – and have established an optimised television broadcasting plan that sees a mix of low-powered transmitters, that operate on the same frequency to improve spectrum efficiency, and satellite reception in the rural areas," he says.

To make a long story short, negotiations are ongoing, with

farmers trying to improve their lot, SKA SA trying to avoid the antagonism of farmers who believe they are being "marginalised and disempowered" and developing solutions for them, and service providers trying to avoid bad press and resolve an intractable situation. At the moment, it would appear that the Department of Science and Technology is smoothing the way with the subsidisation of telecommunications. There appears to be a great deal of trying, but only time will tell if it pays off or if someone has to give.

Internet connectivity for South Africa's SKA bid
Toby Shapshak
When the SKA bid in South Africa was first conceived, it was considered technically impossible. Even if you could raise the money, build the dishes, capture the signals and process all the data, there was simply no way to carry all the information needed, to keep the far-flung dishes aligned and return what they heard from the sky to a single point. Building a dedicated network, just for the SKA, would come at a price, and require ditches to be dug and cables laid through the length of the African continent. And even then, sharing the data with scientists elsewhere in the world would be, at best, very, very difficult.

Historically, Africa's Internet access sucked. Geographic isolation and a relatively small (and therefore not very profitable) Internet market, combined with state-owned telephone companies with legal monopolies, meant a reliance on ageing cables and slow satellite connections. Fortunately, South Africa and the continent has in recent years seen a glut of new undersea cables and intra-national fibre-optic networks. With more of the same to come, data transfer capacity may just catch up to the needs of the SKA, without any need to break the bank. That started to change especially rapidly in 2010, however, and over the next two years international connectivity into Africa increased tenfold, to an estimated 22 Terabits per second

(Tbps), according to Arthur Goldstuck, MD of research group World Wide Worx.

In South Africa, the crucial undersea cable capacity stood at 2.69 Tbps at the end of 2011, with a jump to 11.9 Tbps by the end of 2012, according to the Internet Access in South Africa 2012 study. By 2013, it should have doubled again.[10]

On paper it is only numbers, but for geeks it is a broadband dream, and a powerful enabler of the Internet economy. It also made the SKA realistic. Without the rapid expansion in connectivity, all the radio quiet in the world would not have made South Africa a suitable site for the project.

Until the Seacom cable launched in July 2009, South Africa had only one undersea cable, known as SAT3/SAFE, connecting it the rest of the world. Launched in 2002, it was controlled by Telkom, South Africa's state-owned telecoms company. Even at the time it represented a very small straw through which to access the veritable sea of information that is the Internet, and over the next seven years it proved wildly inadequate.

Fibre optics are a complicated science all on their own, and the way in which increasingly large amounts of data can be pushed down hair-thin lines of fibre is a startling story in itself. But what it means for South Africa and the SKA is quite simple: possibility.

Consider that South Africa had only SAT3's 40 Gigabits per second (Gbps) until Seacom launched, when it was increased to 120 Gbps. That served most of the country's Internet needs, from consumers surfing the Web to large corporations conducting trade. By mid-2012 that total number stood at an estimated 340 Gbps, still too little for the SKA to thrive, but the trend was certainly pointing in the right direction.

---

10 Information contained in this essay was sourced from the following articles: "Internet Access in South Africa 2012" (World Wide Worx); http://manypossibilities.net/2012/04/african-undersea-cable-map-goes-non-linear/ (Steve Song); http://www.techcentral.co.za/infraco-plans-giant-network-overhaul/31772/ (TechCentral); and "The hopeful continent: Africa rising", http://www.economist.com/node/21541015 (*The Economist*).

Also important was the loss of monopoly control of international access, throughout the continent. Where a single state-owned company would see the need for only a single cable, which also represents a single point of failure, the likes of Seacom (with its 1.28 Tbps capacity shared among several countries down Africa's east coast) broke that stranglehold. It also showed others that there was commercial gold to be had in hooking the continent up to the Internet hubs of Europe and North America.

It took a while for the impact to show, especially in terms of the speed consumers could expect and the prices they paid, but before Seacom was a year old, another cable was launched – the Eastern Africa Submarine Cable System (EASsy) in July 2010 – and it provided just less than four times as much as Seacom had, or 4.72 Tbps.

Then followed the West African Cable System (WACS) cable, launched on the eve of the SKA bid finalisation in May 2012 (with 5.12 Tbps capacity). Several others will or could be operational well before the SKA is a reality: the Africa Coast to Europe (ACE) cable, with the same capacity as WACS; the ambitious 12.8 Tpbs BRICS cable intended to connect Brazil, Russia, India, China and South Africa, stretching over 34 000 kilometres; SAex (12.8 Tbps); WASACE, at a massive 40 Tbps; and plans for other connections between Africa and South America.

The driver is profit, not science. Europe and North America have a surfeit of cables serving them, and the developing world is hungry for bandwidth. Cable companies can do the maths as well as anyone, and the developing world has learned the lessons of the likes of India and South Korea, who have built powerhouse Internet economies on the back of fast access.

Within countries the situation is much the same. In South Africa the three major cellular operators (Vodacom, MTN and Cell C), as well as Internet service providers such as Internet Solutions and MWEB have been laying their own fibre-optic networks in a frenzy of activity. These networks are their future, the tools that will allow them to offer cheaper voice services to their customers, as well as the next-generation data services on which their growth will depend. Being behind the curve in providing video streaming and cloud-based computing could be disastrous.

That combines happily with the surge in international connectivity to allow data to flow across and through the country. That data could be funny cat videos, or e-commerce transactions, but could also be the lifeblood of the SKA.

The SKA requires a combination of different networks, says Dr Adrian Tiplady, the South African SKA site bid manager. Dishes within the array's 180-kilometre central core will be on the SKA's private fibre-optic network, cables that run to the dishes and nowhere else. Dishes outside the core, but still within South Africa, will use an existing South African research network. Beyond that, to reach dishes in other countries, those new undersea fibre-optic cables become crucial.

Within the core SKA network, the numbers are staggering, but remote stations still need more than your average home Internet connection. Hundreds of aperture array stations need a dedicated connection capable of between 16 and 32 Tbps, almost as much data as the entire continent will have access to in the first years after the SKA is built. The remote stations, including the 12 in partner countries north of South Africa, will each require 216 Gbps, which is perhaps best visualised as 25 000 standard home connections. This also holds true for the three extra base stations in Ghana, Kenya and Mauritius, which extend the SKA baseline but are beyond what is required for its fundamental functions. That is just enough to get the raw data gathered and to the main data processing site, so a supercomputer can crunch it. Then processed and semi-processed data still needs to be shared with the rest of the world, which will initially be done over a 100 Gbps connection between the primary SKA site and a data centre in Cape Town.

Five years ago, this kind of connectivity anywhere on the continent was unthought of, but now it is not only possible, but a developing reality.

9

The African
VLBI Network

Since bigger is better in radio astronomy, why settle for a collection of telescopes in one place when they can sprawl the globe? This is where Very Long Baseline Interferometry (VLBI) comes in. It sounds intimidatingly technical and a bit of a tongue twister, but the underlying concept is simple: first, a radio telescope is as powerful as its collecting area. The larger your collecting area, the better the telescope. Second, if you have one dish, you can do a limited number of observations. If you have two or more linked together, then things get more interesting. The Square Kilometre Array will be an interferometer that spans Africa and parts of Australia, but the VLBI network effectively spans the globe. All the telescopes in this network operate as one instrument. If you watch a star in Europe, for example, and then also observe it in South Africa, you can work out how far away it is; in some instances, you can even figure out how big it is. But then there's the other side of the coin: if there are radio-emitting sources in the sky that are effectively fixed in position, you can use them to see how tectonic plates on the Earth move because the VLBI telescopes provide reference points, and measure polar motion by comparing the observations of telescopes on different continents. Also, radio telescopes and GPS need an absolute spatial reference frame, which they get through VLBI networks by observing quasars so far out in the universe that they appear as fixed beacons.

To do VLBI astronomical imaging, you need a minimum of four dishes. The United States has its own dedicated VLBI network, but many countries don't have that luxury and their instruments are dual purpose. Also, many countries don't have that many radio telescopes. So South Africa, for example, plugs into the European VLBI network, the EVN.

In the SKA madness, South Africa's VLBI programme – which first began in 1971 with Australia using the modest HartRAO dish – has been overshadowed by the alluring prospect of 3000 dishes. But for decades that programme has been the bread and butter of South African radio astronomy. This is mainly because up until the KAT-7 array came online, HartRAO had the only radio telescope on the continent.

It may be modest in comparison to other radio astronomy observatories around the world, but Dr Nicolson maintains that "Hartebeesthoek reached a level of being a reasonably top-class observatory because it was involved in global programmes of long baseline interferometry. It is equivalent, except in size ... There is no doubt that the observatory is at the cutting edge, certainly in VLBI and VLBI instrumentation, and the capacity to develop and operate those systems."

Imagine this: A group of astronomers from all over the world come together and decide that they want to paint a map of space. There are many astronomers in the United States and in Europe, so their pictures are detailed and accurate. There are a few in Australia and China, filling in their portion of the sky map. But over Africa, there is a fuzzy image, with a tiny picture coming from the southern tip. They can imagine what it looks like over the rest of Africa, but that's what it is: guess work, because they can't construct a fully accurate image. South Africa has played a pivotal role in shedding some light on what would otherwise be pixilated skies, and connecting to these global networks. Take a look at Figure 2 and Figure 3 in the colour section. Imagine that the curved tracks are panels on the surface of a radio telescope – because that's all a collection of radio telescopes, called an interferometer, are. Covering a circle with lots of curved tracks will provide almost all the information needed to make a radio image. Large gaps in the tracks mean the missing information prevents radio astronomers from constructing a fully accurate image.

There are three main VLBI applications in South Africa. The first is celestial detail. As Dr Mike Gaylard says: "When you want to see the fine detail, you have got to spread your telescopes far out." It's similar to having 10 people standing at different positions in a room looking at the same vase of flowers. By having all of these telescopes, it is like looking at them through a pair of binoculars, allowing you to see 10 times the detail. The salient point is that the more radio telescopes you have, and the more perspectives you have, the better you can observe a radio source and the finer the detail you can see.

Then there is astrometry, which is literally a map of the sky and precision measurements of all the things in it. This involves a set of distant radio sources, quasars. They are so far away that they seem fixed, and their positions have been measured very accurately. Using these compass points, you can measure just how far away the nearer radio sources are. Having an array of telescopes means you can pinpoint the exact distance to an object. If thousands of kilometres separate the telescopes, the signal arrives at the antennas at slightly different times because it will take longer to reach a telescope that is further way. The astronomers know the distances between the telescopes, and with the differences in arrival times, they can figure out exactly where the radio-emitting object is in relation to Earth.

Now there is no point in setting up such a sprawling, technically advanced system to get readings that aren't in synch. Timing is key, so each radio telescope is equipped with an atomic clock, which is as cunning and expensive as it sounds – an atomic clock is the most accurate time device on Earth. It is based on the science of atoms, particularly the caesium-133 atom. Atoms have electrons in them, in varying energy states, and when electrons change states, they emit radiation. Since 1967, the radiation emitted by the caesium atom has been used to define a second, with just more than nine billion electron changes in that time. However, while caesium clocks give the most accurate time, they're a bit jumpy, like a thoroughbred racehorse. So radio astronomers use atomic clocks based on hydrogen, which is excited to produce radio emissions at 1420.4 MHz.

So VLBI partnering astronomers send a package of celestial data that is married with a specific time. At a central hub, all of this information is put together. South Africa is part of Europe's network and we send out information to them and, until very recently, this had to be done in hardcopy. Astronomers at HartRAO made their VLBI observations, loaded them onto a (very big) disk – about a terrabyte per day – and sent it on a plane to Europe, and then found out a month later that something had gone wrong with the observation. Thank goodness for the Internet, Dr Gaylard says. "As opposed to recording it all on a disk, putting it on an aeroplane, flying it there and it gets there a week later, now you can correlate it in real time."

One reason that this precision and correlation is so important is because this data forms part of the International Celestial Reference Frame, which is defined using 212 radio sources outside of our galaxy, mainly quasars. This framework standardises astronomical observing, and South Africa has played an indispensable role in this framework since it became a member of global VLBI networks in 1986. But the country's collaboration with Europe in VLBI didn't start with the framework. It began with geodesy, which is the third important reason for South Africa tapping into this larger VLBI array. Geodetic and astrometric VLBI is practised with two receivers at different frequencies, to take out the effects of the Earth's atmosphere. This began in 1986 when South Africa joined Nasa's Crustal Dynamics Program to measure the movements of the continents.

While we mostly think of radio astronomy as humans looking up at the sky and trying to discern and map what is out there, we seldom think of it working the opposite way. If a telescope can pinpoint the absolute location of a celestial object with respect to itself, then it can monitor its own movements with respect to that star or quasar or pulsar. If there are other telescopes involved, you can tell if the object is moving or whether it's just you. Geodesy is about the measurement of the Earth itself, the movement of tectonic plates, the drift of continents.

In fact, Africa's continental drift was first measured using the HartRAO telescope, because – using the international VLBI reference frame – it was possible to determine how far the dish was moving relative to telescopes on other continents. "The movements are very small," Dr Gaylard says. "We are moving northeast at 2.5 centimetres a year."

Dr Gaylard gets quite enthused when he talks about what geodesy is and why the VLBI network is important. He throws his arms out expansively: "We are the reference point for the survey system in South Africa. We measure continental drift; we are critical for geography, measuring polar motion and where all the continents are going; as well as providing an absolute reference for all the GPS stations in southern Africa. This telescope is the absolute reference."

Prof Justin Jonas argues that VLBI is one of the reasons why radio astronomy is such a priority: "HartRAO has played a unique role in radio astronomy around the world because it has this unique position. We are the only big radio telescope currently on the African continent. From here, we can do VLBI experiments with America, with Europe, with Japan and Canada and with Australia. Nobody else can do that, and it's because of where we're located, at the tip of Africa."

But now, after more than 50 years as the only VLBI telescope in Africa, it sounds like HartRAO's 26-metre dish is going to have some imminent competition.

"In order to do VLBI well," says Prof Jonas, "you need antennas, and this dearth of antennas in Africa is a problem. So we always wondered how you would fill Africa with radio telescopes. It's going to cost money. How are you going to do this?" It would appear that Dr Gaylard, armed with Google Earth, has found the solution.

☾

Boredom is an under-appreciated and misunderstood tool in scientific innovation. Albert Einstein was pushing paper in a patent office when he developed his special theory of relativity, which altered the foundations of physics. In 2008, Mike Gaylard, then the associate director for radio astronomy at HartRAO, was bored.

A bearing on the 26-metre dish was broken – 48 years of hard use tends to do that. When, in 1967, new panels were added to the skeletal dish's frame to enable the antenna to work at a higher frequency, the Nasa engineers were satisfied that the bearing would be able to support future planned projects, even with the additional weight. "They did an assessment then to see if it would last through the project, which was their big goal. They figured out how many cycles it would go through, how many turns it does, and 'Yes, it will last through Apollo, no problem,'" says Dr Gaylard.

But in 2008, the bearing finally kicked the bucket. Dr Gaylard keeps repeating "it wasn't meant to be replaced". "The whole polar shaft came out as a 16-ton assembly from the United States; it wasn't designed to be replaced." Since the KAT-7, and later the MeerKAT, was expected to come online relatively soon, the question was whether it should be replaced, should the team at HartRAO build a new telescope, should they bring in people from America to see if it could be fixed? "Eventually it was decided to fix it, mainly because it was the geodetic VLBI reference point for the survey system of South Africa," he says. So engineers were flown in from the United States, specifically from a company called General Dynamics, which specialises in large antennas.

This left Dr Gaylard with time on his hands. So what does an astronomer do when he can't explore the skies? Well, from the sound of it, he became a chair-bound explorer, investigating lesser-known parts of Africa with Google Earth. It is at times like this that you wonder what people did before the Internet.

In the two years that followed, Dr Gaylard sat in front of a computer and hunted for telescopes.

At about the same time as the dish bearing broke, he fortuitously came across a paper by IntelSat, which documented the countries in Africa that had large satellite dishes, and gave a – sometimes very vague – description of where they were.

Back in the old days, before the fibre-optic cables that now flank the coasts of Africa and provide the continent with high-speed Internet, countries used satellite dishes to communicate with the rest of the world, and Telex machines – the modern telegraph, before the days of fax machines – to send data. But the cables made these dishes – and thank heavens the Telex machines – redundant because one fibre-optic cable equals about one thousand of these dishes in terms of its bandwidth.

In the 1970s and 1980s, coinciding with their independence, many African states built these large dishes to transmit voices, TV broadcasts and other data. You needed such a big dish because the satellites had low-powered transmitters. But the march of technology continued: transmitters got bigger, so you could indeed use smaller dishes. Within a decade, these smaller dishes became the flavour of the month, and the big ones were used as backups. Technology continued its relentless advance and now the smaller dishes are the backup, with the population relying on the undersea cables, and the big dishes are collecting dust, rain or being stripped for scrap.

Coastal countries were the first to benefit from the undersea cables, simply because they are literally the first port of call, but some landlocked countries are waiting for the joys of the super speedy Internet to filter inland. They are still using their telecommunications satellites, but for many African countries this outmoded technology is now redundant – a roosting place for birds and a shady stop for local wildlife.

Now, Dr Gaylard knew vaguely where some of these dishes were. Unfortunately, they are usually in remote places, but for convenience – convenience for people who aren't trying to find them on Google Earth, that is – they are listed as being in the closest major city, even though it's a fair distance away.[1] For example, South Africa's original satellite station, next door to HartRAO, is listed on IntelSat's records as being in Pretoria, which is about 60 kilometres away. Being an explorer from a distance sounds as labour intensive and exhausting as wandering remote regions with a backpack and a Swiss Army knife. It isn't just about looking for a place on Google Earth – the locations are vague; you have to do some investigative work and search for old press clippings or any mention of a monster dish. It's strange: you wouldn't think it would be so difficult to find a whopping big white dish in the middle of nowhere.

So you start in the major city, after having looked for any reference to the dish that might help you narrow down its location, and "you work in bigger and bigger circles trying to find

---

[1] A "fair distance" means the kind of distance you wouldn't want to walk, or ride on a bike; unless you're a masochist or your bicycle is fitted with rocket boosters.

these dishes on Google Earth pictures – a needle in a haystack", Dr Gaylard says.

For example, there is a dish somewhere in Zimbabwe, and a doctoral student from Zimbabwe tells Dr Gaylard that she knows about it – it is near a dam where her family used to go for picnics, in a place called Mazoe Valley just north of Harare. Now, "just north of Harare" sounds relatively precise, but it isn't. "So, I followed the road north from Harare through the Mazoe Valley. It's a very lush agricultural area, with irrigation from the dam. I have to go along every road … Eventually, at the end of one road, I see a little white dot – there is a dish sitting there."

But not all African telescope-hunting expeditions are good news stories. There was a Mozambican telescope in Boane. "It turns out there are about five different places called Boane," Dr Gaylard mutters. But someone in HartRAO's technical team said they had passed it on a fishing trip in the 1990s. "So I know it's not far from Maputo and he must have driven past it, but there are actually quite a lot of roads there," he says. "So after I had actually gone through all the wrong Boanes … I followed a road and suddenly there are smaller dishes and then there is a square building with a circular rail track on the roof that is used to run it on, but no dish."

The main concern was that this had happened to most of the dishes in Africa, but that doesn't seem to be the case. Dr Gaylard has rediscovered, uncovered and tracked down 26 large dishes in 19 countries to date, and there might be more. In terms of SKA partner countries, South Africa has three, and Ghana, Kenya, Madagascar and Zambia have one apiece. But other countries also have this wealth of antennas: Nigeria has three; Algeria, Cameroon and Egypt each have two; and Benin, The Democratic Republic of Congo, The Republic of the Congo, Ethiopia, Malawi, Morocco, Niger, Senegal, Tunisia, Uganda and Zimbabwe all have one.

There is great interest in setting up radio astronomy programmes in African countries, even those that aren't SKA partner countries. For example, two Nigerian astronomers at the University of Nigeria in Nsukka did their doctorates at HartRAO and have been pushing to have radio telescopes built in their country. Unfortunately, the cost of building a radio telescope from scratch is daunting. But from Dr Gaylard's research he discovered that there were actually a number of disused dishes in Nigeria – he could even measure their size using Google Earth. So, the once-upon-a-time HartRAO students went to one of the sites, and were met by a guard. When asked if they could speak to the people inside the facility, they were met with, "I am the people". Inside the facility it was allegedly like the *Mary Celeste*, the infamous ship that was discovered abandoned in the Atlantic Ocean in 1872. With the exception of one missing lifeboat, everything was still on the boat, valuables, food, and more than 1700 barrels of booze destined for Italy. Only the people were missing, and the same could be said for one of Nigeria's old satellite dishes: "The equipment is all there," Dr Gaylard says. "You turn the lights on, the PCs, the oscilloscopes, all the equipment and machinery is there. It had just been switched off."

One thing is for certain, if the bearing on the HartRAO hadn't broken, none of these discoveries would have taken place – running an operational radio observatory is a rather time-consuming activity. There definitely wouldn't have been time to telescope hunt through Africa. But the main hero, according to Dr Gaylard, is Google. "I would find what has been in the press and what Google has found, and then what Google's satellite shows. The African VLBI Network would not even be a concept if it wasn't for Google."

The African VLBI Network is arguably more exciting than the SKA. Yes, the SKA will bring prestige and cutting-edge science, but the network of African radio telescopes is something that is Africa driven, and a push that comes from within the continent.

It will definitely have benefits for the SKA, but it is independent and hopefully won't take 20 years to get going. "When the SKA comes to Africa, life is going to be a great deal easier if we have people who know radio telescopes and how to do radio astronomy in those [partner] countries," Dr Gaylard says.

It also helps if there is already a big dish there, standing idle. While it costs money to convert satellite dishes into radio astronomy telescopes – about R10–R15 million – it is much cheaper than starting from scratch. But there is a more important reason: radio astronomy has been held up as a way to increase innovation and drive skills development. Why should this chance be confined to South Africa, when other African states want to get in on the action? The discipline is recognised as a great way to develop highly skilled people, which is something that the continent really needs.

This isn't just an African trend. There are countries around the world sitting in a similar situation. For example, Russia has a number of redundant military satellite stations that they are converting into radio telescopes. Latvia and Lithuania are also turning their abandoned big dishes into astronomy instruments. Even on the other side of the globe, Peru has an old satellite dish in the Andes, about 3000 metres above sea level, which they are turning into a radio telescope. New Zealand (Warkworth), England (Goonhilly) and Ireland (Elfordtown) are doing the same. Converted satellite antennas are already operational in Australia (Ceduna) and Japan (Yamaguchi, Ibaraki).

The most prominent success story in Africa is Ghana. An interesting aside about this small country in the western bulge of the continent: it was the first sub-Saharan African country to achieve independence from its coloniser, the United Kingdom, in 1957 and is one of the largest producers of cocoa, used in chocolate. As far as the author is concerned, its importance in chocolate production makes it a vital little country. It also has an old 32-metre satellite dish and a smaller 16-metre one that still carries some satellite traffic at the Kuntunse Satellite

Earth Station outside Accra. Now, in most African countries, including South Africa, the state originally built and operated these antennas, and communications were the prerogative of the country's government. In South Africa, prior to 1991 there was only the Department of Posts and Telecommunications, which handled all of the country's communications, including the postal service and telephone system. Telkom, which is now Africa's largest integrated communications company, was established in October 1991, along with the Post Office. Privatisation of communications appears to have been the trend, with the same thing happening in Ghana. Vodafone owns the antennas on the Kuntunse site, but the larger dish has been out of commission for a decade, just sitting there taking up space.

Vodafone has agreed to donate the 32-metre dish to the Ghanaian government for radio astronomy. At the moment, everyone seems to be negotiating the red tape of acquiring the dish, which should be an operational telescope by the end of 2012. Prof Justin Jonas appears rather matter-of-fact about the possibility of a new telescope: "I've been over there myself, and we've been working on building a receiver for that dish, working closely with the local Ghanaians. They're setting up a radio astronomy institute there." But Dr Gaylard is much more outwardly enthused. A big smile stretches across his face as he starts talking about it: "I knew from satellite photos that there was a 32-metre [dish] on top of the roof, and it is still there! You can see it on the photos. So we get there and say, 'Well, are you using this, the 32-metre?' 'No, no, we shut that down 10 years ago.'"

It would be wrong to call sophisticated radio astronomy equipment a toy, but Dr Gaylard looks like a child who has just opened a Christmas present, after his parents had convinced him that Santa had decided not to come to his house that year. His hazel eyes twinkle excitedly and his hands become a blur of gestures. "So we climbed all over it. The thing hadn't moved in 10 years, but they had been doing basic maintenance on it –

painting it and so on. Basically, there were layers of paint just building up, but that's good because it preserves it."

Although the paint meant they couldn't start driving the telescope immediately, within a day South African radio astronomers had photocopied the manuals and had a good idea about what sort of telescope they were dealing with. "We went into the control room. The equipment was all still there, but that's it. It wasn't being used." Hopefully by the end of 2012 that control room will be filled with radio astronomers, making the first observations on the continent's first radio telescope that isn't in South Africa.

Perhaps the most important thing to note is that this is not a South African project. Yes, South Africa has the expertise and will seed the science, but radio telescopes in other African countries will belong to those countries. "These countries will have their own radio telescopes that they own," Dr Gaylard explains. "We won't own them. We will facilitate them, but it will be their radio telescope."

What these countries do from there is their prerogative. The first step would probably be to join the European VLBI network, as South Africa did. But if another two satellite dishes can be converted into radio telescopes – which seems like a realistic possibility – then Africa could have its own VLBI array. Since Nasa first decided to establish a spacecraft tracking station at Hartbeeshoek, South Africa has been the only radio astronomy power in Africa, the only fish in the pond. Because of that, many students from other African countries have done their postgraduate degrees in radio astronomy – be it in pure astronomy, engineering, software development – in South Africa. The number of students who have done work at the facility has increased significantly with the SKA Project Office's human capital development programme. But now these countries will have the option of doing their own training, and radio astronomers who trained in South Africa could possibly have radio telescopes of their own. "They can train students, they can get into receiver development, software, all sorts of hi-tech stuff that we can do here, they can do," says Dr Gaylard.

In June 2012, the African Renaissance Fund approved R120 million to go towards the African VLBI Network, and what makes this so exciting – possibly even more than the SKA, even though it will probably sink into obscurity next to the dazzling leviathan array – is that it is a project that Africans can call their own.

> **The origin of death**
> The hare, with his long ears and split lip, was not always as we now know him. Once he was a man, and it is because of him that the Moon cursed us all to die.
>
> The Moon and the male ostrich are the only creatures in the world that can die and come back to life. The ostrich can be resurrected with a feather, but the Moon – the Moon contains in his glowing belly the secrets of life and death. No one knows as much about treading the twilight between dying and rebirth as the Moon, and the hare should have known that and should not have doubted him.
>
> A long time ago, a man sat crying – his mother lay upon the cold floor, her eyes closed and she would not wake. The Moon, his stomach thickening, was rising in the sky, and he saw the man and tried to comfort him.
>
> "Do not cry, your mother will rise, as I rise. Dry your tears, you do not need them," he said.
>
> But the man would not listen because he had seen his mother still and lifeless. Again the Moon told the man that all would be well.
>
> And the man, his faced swollen with tears, shouted at the Moon, "Do not tell me to stop crying. My mother – she lies right there in front of you – is

dead. She will not live again, and I will cry all the tears out of my heart."

Luckily for the man, the Moon was weak and clawing his way back into life, otherwise he might have killed him. Instead, in his anger, the Moon reached down from the skies and smacked the man in the mouth with his fist, splitting his lip.

As the man clutched his severed and deformed face, the Moon shouted: "This man shall bear this scar to the end of time. I, who die and return living, told him that his mother would return to life, and he contradicted me. So he will become a hare."

With that, the man felt his bones shrink under his skin, which became itchy as hairs pushed through its surface. And as his ears lengthened, the Moon's roar became deafening. "He shall always bear that scar to show people what happens when they insult me. He shall spring away, and the dogs shall chase him. When they have caught him, they will tear him to pieces and then he shall truly die. And so too will all people: unlike me, when they die, they will not rise again."

But the Moon was still riding the crest of his rage. He cursed the hare, not only to die, but also that he may lie on the ground and be covered in vermin. He would not be allowed to seek the safety of the bushes, but must sleep on the open ground where the ants and fleas could crawl into his fur and nestle behind his ears. And so, with his grimace-deformed face, the hare frantically shook his head from side to side, tormented and crazed as he felt the creatures wiggle a path up his body.

So you must respect the Moon, who is quick to anger, and when you kill a hare, you must always cut out a part of the meat and leave it for the Moon – because that is human flesh, the only remainder of what the hare once was.

10
Benefits

No telescope – no matter how big it is – is going to solve a country's problems, or even those of the small town closest to it. Small towns in South Africa are microcosms of greater societal problems endemic in the country. And perhaps it's a sense of skewed self-importance, but those problems seem to be especially insurmountable in rural South Africa.

The key word in South Africa at the moment is jobs. With an unemployment rate lurching between 20% and 25% for the last 10 years, job creation drives South Africa's economic policy. There is also an electorate whose main concern is the availability of jobs, and political parties know that their chances hinge on convincing voters that they will be able to deliver on the elusive job-creation holy grail. Housing, education, potable water, affordable electricity are what your average South African wants, and all of this is synthesised into one word: a job.

It is difficult to justify exorbitant expenditure on scientific equipment when the reality is that children need to be schooled and people need roofs over their heads, not to mention the expanding social grants system that keeps the country from anarchy. Large scientific instruments are seen as a fix-all, not just by politicians trying to drum up support, but by the people who actually live in these towns – a misguided perception that a consortium of science institutions will be able to deliver where their government has failed.

☾

"We're lucky" is the refrain from the community of Sutherland in the Northern Cape, home to the optical SALT. "We're lucky to have the telescope." Some people argue that a developing country, such as South Africa, should host large scientific instruments because it would create employment and boost human capital development. Others say that South Africa should focus on poverty alleviation and the problems facing the country, rather than spend millions on arcane scientific equipment.

The Square Kilometre Array is similar to SALT in a number of ways: the projects are funded by international consortia, they will both be based in small farming communities in the Northern Cape and will focus on pure, experimental science. The question then is whether large international scientific projects actually benefit the communities in which they are based. According to people in Sutherland – from headmasters to small business owners to police officers to waitresses – the answer is an unequivocal yes.

With a population of less than 4000 inhabitants, Sutherland relies on tourism and agriculture to generate income. Unfortunately, agriculture is by its very nature seasonal, offering unreliable employment when demand warrants it – whether it is potato harvesting or sheep shearing. As the employment opportunities fluctuate, the town's unemployment rate sits between 70% and 75%, and many are reliant on social grants.

Despite its bipolar temperatures, with sweltering heat in summer and freezing misery in winter, you go to Sutherland to look at the stars. There are very few places in South Africa that have such a panoramic and undisturbed view of the Milky Way, which looks like a glowing dust trail scattered through the sky. If the sky was any less awe-inspiring, you'd be disappointed – after all, Sutherland was chosen as the site for SALT because of its star-strewn ceiling.

The town's main draw card is stargazing, and although activities have been added to the to-do list, such as hikes and indigenous plant trails, these are more deal sweeteners than the main event. Tourism offers stable employment for many in the town – whether as small business owners, tour guides or service and maintenance staff.

For the money spent on SALT – about a $20-million infrastructure spend, with an annual operations bill of about R20 million – it isn't a big facility, and there are only so many low-skilled jobs it can offer to the mainly unskilled town of Sutherland. The main benefit for the community has been downstream and indirect. Domestic and international interest in seeing the stars has bolstered the local hospitality industry, with coffee shops and bed-and-breakfasts mushrooming up around the town and dotting the countryside. Small business owner Jurg Wagener, a former bank employee who decided the hospitality industry was a great place to spend retirement, owns one of the 45 guesthouses and guest farms in Sutherland. "I often have bus loads of tourists, people coming throughout the year," he says proudly. He employs 10 people at his guesthouse, although he admits that most decisions have to be signed off by his "boss", otherwise known as his wife.

Like all small towns in South Africa, if you want to find the central hub, just ask for "Church Street". The epicentre of the town dates back to 1899, and was one of the sites for the numerous skirmishes that marked the South African War. Now, Sutherland has a long memory, and no tour of the town would be complete without a full recounting of how many people there were at any given time, where they were stationed and how they dared to carve their names on the church walls.

The town's main road – dusty and wide enough to allow two carts to pass each other – is flanked by guesthouses, curio shops and restaurants, each trying to catch the attention of passing stargazers. And there is plenty of attention to be caught: according to South African Astronomical Observatory officer Anthony Mietas, SALT has more than 10 000 visitors a year. Now, that's a rather hefty guest-log for any small town, and more than double the number of Sutherland inhabitants.

Three golden words are one way to justify investing in a large international scientific instrument in a poor country: human capital development. And in many respects it is a fair justification. South Africans are more likely to be exposed to world-class astronomy by having an instrument such as a SALT; universities tailor their science and engineering programmes towards such instruments because it's possible to get firsthand experience on excellent equipment. But the reality is that very few – if any – of the town's inhabitants will become astronomers or even be directly employed by the facility. The major spin-off is that no country, in good conscience, can have a multimillion-dollar science observatory looking down on a school that is falling apart.

So one of the main benefits is that Sutherland's local schools, and the hostel, which houses 85 pupils between Grade 1 and Grade 12, have been upgraded through funding from the National Research Foundation and have been labelled Dinaledi schools – schools with a special focus on Maths and Science. However, the schools would not have received such attention from the government – because there are many schools that are falling apart, without textbooks and too few teachers to too many pupils – had it not been for the presence of the telescope so close to the town.

Primary-school headmaster Neville van Wyk is a middle-aged man with a round belly, and as he walks into the classroom, the children fall silent and pretend they hadn't been talking loudly just moments before. You can see from the way that he talks about the school and its learners, with passion and authority, that he cares about what happens within those walls.

He says that the high school has had a 100% matric pass rate for the past 10 years,[1] although the classes have been very small, with only 18 pupils matriculating in 2010. He speaks a great deal about the reasons why so few children get to matric level: seasonal workers who take their children out of school when they go and work on faraway farms and a high drop-out rate, either from a need to start working early in life to support a family, the immediacy of which cancels out future earnings a matric might bring, or teen pregnancy. In fact, he cites teen pregnancy as one of the main problems, and links it to rampant alcohol abuse. On the day when government grants, including pensions, are disbursed, you can witness firsthand the extent of the problem: the main street of Sutherland is swarming with people, most of whom are congregating around the door of the bottle store. Many drink openly on the street corners, some lurch down the road, and one man even thinks that a hard pavement is the perfect place to take a nap.

Warrant Officer Marius Malan cites alcoholism as the most serious problem facing the town, and the cause of most of Sutherland's crime. "But compared to other places, it's very good. It's mainly domestic violence. People withdraw the charges when they're sober … We have a murder once in a blue moon. We've only had one this year [2011]," he says, adding that it was also related to alcohol abuse.

---

1 Once again, this is another South African-ism. Matric is a school learner's last year in secondary school, and their final exams in their matric year determine whether they qualify for university acceptance.

Also, although poverty is a problem and a lot of people in the town are poor, no one is starving, he says. You can see that from a drive through the town. The socio-economic divide is obvious. The main road is fairly well kept and the houses only extend about two blocks away from the road. As you continue through the town, towards Williston, you begin to see it. On the left, neat rows of painted government-built houses, each with its own very small garden, are built on the sandy land. On the right, many of the houses are built of wattle and daub, or some non-cement equivalent. Even driving through the area on the right of the road, there is only one house that is truly decrepit. Despite the obvious poverty, people get by, which is more than can be said for other small farming towns.

The bottom line is that without the telescope, Sutherland would struggle like many other small farming towns in South Africa, where employment opportunities are thin on the ground and future prospects even slimmer. Its proximity offers more than an abstract hope that our country will become more scientifically advanced and attract more foreign direct investment: the trickle-down effect equates to jobs, improved education facilities and social development on the ground, which is what small towns really need.

☾

There is a stark difference between Sutherland and Carnarvon. While Sutherland definitely has its problems, there is a sense of optimism. The telescope appears to have lent the town an abstract purpose, while Carnarvon is caught up in slow-moving entropy – slow enough that people don't galvanise into action, and real enough that a tang of despair permeates the air. Disclaimer: while this may not be obvious to inhabitants, it is the first thing an outsider notices. By 2011 Carnarvon was in a sorry state, with jobs few and far between, and easily accessible alcohol compounding the problem and the main cause of domestic abuse and crime.

The easiest way to get from Sutherland to Carnarvon is via the R354 road to Williston, which could callously be called the town that didn't get the SKA. Since the road has a name and can be found on GPS or Google Maps, you may be fooled into thinking that the R354 is a road in the true sense of the word, with tarmac and painted lines. This would be a mistake. While the drive is picturesque, it is a 137-kilometre dirt road and it makes up for what it lacks in cellphone reception in potholes and rocks. It is best to let someone know when you're leaving one town and expect to arrive at the other: if a pothole takes out two tyres – which, although unlucky, is not entirely unlikely – you're in for a long, hot walk.

Once you hit Williston and a tarred road, you may be overcome with excitement and begin speeding. This is ill-advised, because the few drivers on the R63 road to Carnarvon have a similar *joie de vivre* or are just homicidal maniacs. There are few places in the world where a 16-wheeler truck can over take you at 160 kilometres per hour.

An initial impression of Carnarvon can be summarised in two words: hot and dusty – although to be fair this was in spring. In winter, it could be described as cold and dusty. You can see that the town itself hasn't changed much since German missionaries first set up a Rhenish Mission station in the area. In fact, many of the buildings, which date back to the 1800s, bear testimony to this missionary zeal. The church, which is the focal point of the town's centre, was built in 1858, and is encircled by wide tarred roads with dusty shoulders. A possible reason for the dust is that the town is built on a plateau and surrounded by hills, which shrug sand onto the flat land. It could also be because this is the Northern Cape, where sand is the staple and bushy shrubs a sporadic adornment. While "permeating despair" might be considered artistic exaggeration by the uncritical or people who live in Carnarvon, the figures back it up.[2]

If you live in Carnarvon, chances are you were born in Carnarvon – about 60% of the locals – and about 90% of its inhabitants want to be buried there. Many people choose to live in small towns, but this is usually in the upper-income bracket. Often city-goers relocate to small towns in search of a less frenzied existence, but poorer people have no other choice, particularly the women.

The area has seen increased migration since the 1990s, with people leaving rural areas for the urban towns, and many leaving the location all together. The main reasons cited are the job hunt (42.9%), being closer to employment (21.4%) and proximity to education (14.3%). Also, the town has a disproportionate number of female inhabitants and an increasing number of women-led households.

From this, you can deduce a few uncomfortable realities. First, men are more mobile; societally, there is pressure on them to have gainful employment, and because of the dearth of economic opportunity in Carnarvon they have to travel to other places to search for jobs. The skewed gender demographic – with older women being one of the largest population groups – indicates that when men leave, they do not always come back, resulting in an increase in women-led households. It also indicates poverty: in a single-mother household there is often only one income supporting the family, and that income is less than it would be

---

[2] The following figures are taken from a report by the University of the Free State's Centre for Development Support. It is a follow-up study done in 2009, with the first having been undertaken in 2007. The SKA Project Office commissioned the study to compare the effects that the MeerKAT and SKA projects would have on the surrounding towns. Naturally, Carnarvon is included, as well as Williston and Victoria West. Williston is likely to enjoy trickle-down benefits from the telescope, although not quite the torrent expected in Carnarvon. Victoria West is – unluckily for it – the town that got the placebo medication, and is used as a metric to compare the SKA benefits, while it is expected to stay pretty much the same.

if the household was led by a man simply because women are paid less.

Any analysis of Carnarvon must tattoo the real racial divide in the town across its forehead.[3] From the area in which inhabitants live to the amount they earn, the division falls along racial lines, with white people living in more affluent areas and earning more money. Unlike many other places in South Africa, the majority of the population is coloured[4], many of whom are employed in agriculture. However, agricultural incomes are the lowest in the area, compared to non-agricultural wages, with the notable exception of white farmers who earn substantially more than most people. The inequality between the races is marked, and a source of ongoing tension. It also means that when a project such as the MeerKAT or SKA comes to town, the first question is: How is it going to benefit the community? And the second: Why are white people benefiting more than everyone else?

There are a number of reasons for the lack of opportunity in Carnarvon, the primary one being that the town is remote and lacks the economies of scale experienced by other more industrialised places. This lack of scale infiltrates all aspects of this micro-economy: it isn't big enough to be an attractive market for most large suppliers; the infrastructure is skeletal and transport a major cost for anyone wanting to travel further afield. The infrastructure issue links into a larger constraint: such a small population is not heard by policy-makers. While there are politically active factions, these inconsequential areas are mainly

---

3 Notably, the Centre for Development Support survey fails to mention the racial split in Carnarvon, Williston and Victoria West, a rather large and glaring omission. Consequently, all data concerning racial demographics is taken from the PROVIDE project's "A Profile of the Northern Cape Province: Demographics, Poverty, Income, Inequality and Unemployment from 2000 until 2007", which was published in 2009. The national and provincial departments of agriculture are the stakeholders in PROVIDE.

4 To my international readers, take a deep breath: this is not a racial slur. In South Africa, "coloured" (of mixed race) is a distinct racial and cultural group, and is accepted as an autonomous demographic.

ignored – unless it is election time, when manifold promises are made and later forgotten by the winning party. Such a small demographic in a remote area lacks the political clout a similar-sized group would have in Gauteng, for example.

The town relies on agriculture to generate revenue. Nationally the sector's labour-absorptive ability has been declining since the 1990s, so this problem isn't confined to the Northern Cape. In the 1990s, agriculture was relatively protected and benefited from extensive government subsidies. Since the new democratic dispensation, when South Africa became open to other world markets, agriculture has faired poorly because it no longer has the subsidies to make it competitive. If that wasn't bad enough, technology began kicking agriculture when it was down, at least for farm workers. Increased mechanisation – which is necessary for South African farmers to compete with those in other countries – meant that fewer manual labourers were needed because machines could do their jobs. This explains why, although there has been increased – although not impressive – growth in the agriculture sector, it can no longer be an employment sponge, absorbing large numbers of people.

This has led to people leaving towns such as Carnarvon to search for jobs further afield. But while there are large numbers leaving the rural and urban boundaries of Carnarvon, the people who stay are moving from farms to urban areas – which in a town like Carnarvon is problematic because there are very few jobs in town, and even fewer that are available.

Part of the reason for this is the Extension of Security of Tenure Act, which was passed in 1997. This is a perfect example of the way to hell being paved with good intentions. This Act was passed to secure farm workers' residency on farmers, and meant that a worker could not be unfairly evicted. For example, a 60-year-old-plus worker who had worked on a farm for more than 10 years and could no longer work because of disability or sickness could not be evicted. However, while the aim was to protect workers' rights, it effectively meant that farmers stopped inviting workers to live on the farms because they would not be able to evict workers without following bureaucratic and often drawn-out proceedings.

Perhaps the most startling figure contained in the report by the Centre for Development Support is that about 80% of the Carnarvon population does not have a matric. Only people who were older than school-going age were included, so there is some hope for the next generation, but with such an overwhelming majority having no form of qualification, economic development and job creation are only buzz words and are little more than a pipe dream.

Inhabitants are now trying to leverage the MeerKAT to improve their lot, but things such as education and infrastructure are not the responsibility of a scientific project.

But, at the same time, government grants keep the wolf from the door in many a Carnarvon household. The Northern Cape has the lowest recorded gross domestic product of South Africa's nine provinces, and its unemployment rate is higher than the national average, which sits somewhere between 20% and 25%. These grants – mainly old-age pension, disability and child support grants – are the economic backbone of the town, and without them people would starve. Perhaps the most pointed indication of poverty is the expenditure priorities of most people in Carnarvon: first is food, second is electricity and water, third is debt. However, with such extensive inequality in the town, there is a minority for whom these things are not top of the list, particularly in the white community.

Wilfred Horne is a detective with the Carnarvon branch of the South African Police Service and the chairperson of the Carnarvon Stakeholders Forum, which negotiates with the SKA Project Office. He sits upright in his chair, and chooses his words – which are overlaid with a thick Northern Cape coloured accent – carefully. He tells you upfront that he has been involved in political structures his whole life, notably the African National Congress Youth League, but even if he hadn't volunteered the information you would be able to tell: for Horne, everything comes down to politics, and he speaks in the meandering metaphors characteristic of someone who is used to political platforms.

"It is difficult to avoid engaging in politics, there is politics wherever you go," is the first thing he says, adding that he means party politics and racial politics. "In Carnarvon, we have these parallel divisions between the racial groups, and the SKA has to bring these two groups together."

Lavona van Schalkwyk is a short woman, with a motherly demeanour and surprisingly red nail polish, and she is also a primary school teacher and Horne's deputy in the forum. The first question raised at the initial meeting of the Carnarvon Stakeholders Forum, she tells me, was: "What is it going to give back to us?" "People expect work, that the unemployment issue will be addressed, that they will be able to buy a house and provide for their families," she says.

This might be the greatest issue for the SKA Project Office: expectation management. In a town where people have few prospects, where even if you manage to get a matric there isn't work for you, you will latch onto any opportunity to raise your fortunes. And it's a difficult situation for everyone involved. The SKA SA Project Office is a scientific bureaucracy – its main job is to get the MeerKAT up and running, and then to be involved in the construction of the SKA. It would be unconscionable to build a world-class radio telescope, and then do nothing to help the community, but at the same time, the telescope cannot fulfil the function of government. While jobs may be created through initial infrastructure build and a larger tourism office, as well as an expanded scope for the tourism and hospitality industry, it cannot "save" Carnarvon.

SKA SA director Bernie Fanaroff says the project has always been upfront on what it will be able to deliver, but that inhabitants will continue to hope: "We've always been aware that irrespective of what we say people will develop expectations because in a very poor society anyone who has resources is going to be looked at to make things happen."

But the SKA SA Project Office has only limited resources. It can help with education, although it cannot fix the troubled, and some might even say broken, education system. It can offer a few jobs, but cannot provide employment for the whole town. Dr Fanaroff says that this hasn't been confined to the poorer residents, but also wealthy farmers. Telecommunications are a problem in the area – Telkom lines have fallen into disrepair and, as we have noted, it isn't economically viable to fix them. Cellular providers use the same argument for why they haven't extended their services to the remote and sparsely inhabited region. This situation has been continuing for a number of years, but the arrival of the MeerKAT project – with its radio frequency specifications – means that there is an outsider to blame, who is then expected to fix the situation.

"We help where we can," Dr Fanaroff says, "but we can't solve the area's problems. We can't deliver services. We don't have the resources."

The follow-on question is whether SKA SA should be expected to take on the role of the government, and provide the basic services that these people need but have not been given. But similarly how can people steeped in poverty watch all this money flowing through their town and not want to get in on the action? This is especially emphasised in a town like Carnarvon, where there is little else that is likely to raise its inhabitants' fortunes.

When the MeerKAT and the SKA fail to deliver on these unrealistic expectations – from racial harmony to a surfeit of jobs – what will happen then? Already townsfolk are beginning to have negative perceptions of the telescope. Horne hits the nail on the head when he says: "When we heard about the SKA, everyone thought 'This is a solution to the problems I have'. But with the dragging of development, people are beginning to think there's nothing in it for them. This is our big challenge [as the Carnarvon Stakeholders Forum] because there are such big poverty margins in Carnarvon."

A definite downside of the SKA project is that the chances of people being able to buy houses are becoming increasingly unlikely. While locals may begin earning more, they will not earn enough to keep up with the property price spike. The possibility of the SKA, and the accompanying international interest in this small town, resulted in a perceived surge in demand for houses, even though there was only a minimal increase in demand. Driven by this perception, property prices began to rise, effectively putting houses out of reach for Carnarvon residents.

But there is another more pervasive concern that no report or survey on the area will tell you about. People are resentful, to the point of being envious, of those who manage to make a better life for themselves. It may be the psychology of a small town, where everyone knows everyone else's business; maybe it is because many people have so little; but in Carnarvon, there is an undercurrent of resentment. This resentment, for the most part, hinges on relative inequality. This could be because many people in Carnarvon are effectively "trapped" there and, consequently, the people with whom they can compare themselves are the people they see every day. So there is an internal resentment of people who are similar to you, but have managed to do better. They are your main comparator. The next level is a racial one: when someone of a different race is earning more than you, or has improved their situation, while you have not. For example, one person said: "Whites are benefiting and advancing from the SKA enrichment. A non-white with the same qualifications is not getting the same opportunities."

This perception isn't without basis, and is a hangover of apartheid days. Some 46 years of legislated racism means that white people in Carnarvon still hold the majority of the capital. One of the main beneficiaries of the project to date has been the hospitality industry, because visiting site managers, scientists, engineers and infrastructure-related personnel have been staying in the town while they work on the MeerKAT site. They stay in bed-and-breakfasts, eat at the restaurants and spend money in the town. However, the majority of accommodation outlets are white-owned, meaning that, in this instance, a minority is directly benefiting from the SKA project, whereas others aren't. It should be noted that increased business tourism means that the owners of these establishments hire more people, buy supplies from local business owners, and that there is a trickle-down effect. But the perception is that, once again, a white minority is benefiting.

Dr Fanaroff acknowledges the imbalance of benefits: "There's no doubt that the area has seen an upturn in business, but that upturn hasn't been evenly shared among all parts of the population."

So, in many instances, this resentment may be justified, but it is so widespread that it is difficult to believe that everyone in the town has been short-changed. There is one issue in particular that comes up when you speak to people about the SKA – well at least in 2011, when the subject was very topical. As part of its outreach project, the SKA Project Office had chosen a number of students to study at a Further Education Training college in Kimberley, through which they could become proficient in a trade. No one in the town simply said: it is great that children from Carnarvon have been given an opportunity to better themselves. First, it was not enough. There should have been more learners chosen. Second, the wrong children were chosen. A local source, who did not want to be named, said that two of the learners attending the college were drug addicts and didn't deserve the chance. You get the impression that people believe that if everyone can't advance up the social ladder then no one should.

At the same time, Carnarvon is a town of broken promises. Political parties have come into this town, promised locals the world in exchange for votes, used them for the power they can confer, and then left nothing behind but disillusioned hopes.

Horne says: "People come and make promises and don't

come through." He asserts that they ascend to positions of power and then not only forget the people they serve but also, being hungry to retain their job, with its associated perks, they focus their energy on consolidating their position and often turn a blind eye to transgressions, instead of addressing Carnarvon's manifold problems. This morphs into a slide of patronage in which leaders protect their cabal and create opportunities for undeserving supporters.

Consequently, suspicion and resentment infiltrate every tier of the town. But despite the audible whispers of discontent, many believe they are lucky that the telescope will be built there. The most resounding nod of approval was that the project had literally put Carnarvon on the map, the weather map that is – the national television broadcaster now includes Carnarvon on its nightly weather bulletin.

Van Schalkwyk is particularly optimistic, especially about the benefits for the town's children – although this isn't particularly surprising since she is a schoolteacher. "The SKA is doing a lot, at least for our kids. The most important thing is that our kids are going to benefit, and we need to bring that message to all our leaders ... We should not be focusing on what we should get."

While Carnarvon is upset that it isn't getting more, Williston just feels generally hard done by. On paper, the two towns are interchangeable: the same problems, the same demographics, similar economies. But on the most important piece of paper in this instance – a map – there are 143 kilometres of road separating them. So while Carnarvon is 10 kilometres away from the turn-off to the MeerKAT site, Williston is 133 kilometres, making the latter the runner-up. With the theme of relative resentment now extending between towns, Williston believes that Carnarvon has been unfairly privileged, and thinks it also deserves to benefit from the large radio astronomy project. Carnarvon inhabitants think that the SKA should be doing more to improve their situation, but are still excited that *something* is happening. Conversely, Williston residents are filled with animosity that their town was overlooked. The report by the Centre for Development Support makes an important observation: "It is a matter of time before political interests move in to take advantage of the situation." This is a fair concern because the dashed ambitions of Williston allow for a common enemy – and one that is not elected officials who have forgotten all the promises they made to get there. It is much easier to demonise an outside organisation, and feel slighted, while ignoring geographical facts. But, at the same time, can you blame them? An unexpected ticket to increased prosperity was handed to their next-door neighbour and not them.

For the small towns near large science projects, it isn't about the science. It isn't about the technological advancements. It isn't about increased foreign spending. What do they care if €2.23 billion is being spent on the project if it does nothing to improve their lives? But if the experience of Sutherland is anything to go by, there is light on the horizon for Carnarvon. If you look at what SALT has done for Sutherland, it is difficult to imagine what a project like the SKA will do for towns in the Northern Cape. Already, there are direct benefits from SKA SA, such as a computer lab in the school in Carnarvon, training for locals to become artisans and the deployment of Maths and Science teachers, but more benefits are coming – this area will become an international hub for radio astronomy. About $20 million was spent on SALT, but when we talk about the SKA, we're talking about billions of dollars. When Phase 1 of the SKA comes online in 2019, it is very likely that it will be difficult to recognise Carnarvon, and surrounding towns such as Williston – and the difference will be even more marked, considering the base from which this development is starting.

11
SKA science

There are three important-to-humankind reasons for building large scientific instruments. Of course, there are the benefits that trickle down to the host country, international collaboration and those sorts of things, but you can't use reasons such as these to motivate for billions of euros. No, your justifications have to be of groundbreaking intergenerational importance. As a child you wouldn't have told a parent you needed R100 to make yourself sick on sweets – you needed the money to stem starvation in India, or develop an amateur telescope to search for life on other planets. You had to think big.

So, there are three important reasons for the Square Kilometre Array. The first is what scientists want to use it for and, although the plans may be grandiose, they have to be backed by a strong scientific case ... rather than a bald-faced lie that not only got you into trouble, but also didn't get you the R100 you wanted.

The second is technological innovation. Large science projects are ambitious, and they think 20 years into the future. To envision what sort of science they will be able to do in 20 years, engineers and designers typically look to Moore's Law. This is a fundamental concept in computing hardware. Technically, Moore's Law relates specifically to transistors, semiconductors that are present in all our modern-day electronics, but it can be extrapolated to other aspects of computing and technology. Moore's Law states that the number of transistors that can be placed in a certain space on an integrated circuit will double every 18 months with the cost staying the same – although Gordon Moore later argued he did not say 18 months, but two years. However, it was too late, and Moore's Law had been established. More transistors on a circuit means that it can perform its task faster because it has more processing power. The underlying point is that technology gets faster and cheaper as time goes by.

Consequently, a lot of the processes that will be used for the SKA don't exist yet, but in 2024 when the mammoth telescope is fully operational, they will indeed be available. There is the complicating factor that large projects like this speed innovation along. For example, the SKA has partnered with a number of leading industry players to develop the technology of this seemingly distant future.

The third, and perhaps most interesting reason, concerns the things you don't expect to find. A great deal of science is stumbled upon, rather than sought. Serendipity – a "happy accident" – has a special pride of place at the table of science. Although it is recognised that you can't win the lottery without buying a ticket – and that accidentally making a groundbreaking discovery requires scientific acumen and equipment – many breakthroughs occurred when scientists were looking for something else.

Penicillin is perhaps the most obvious and well-known

example. Sir Alexander Fleming received a Nobel Prize in Medicine in 1945 as a result of a mistake, which has subsequently saved millions of lives. During the First World War the most common way to die, aside from metal entering your body or the loss of important parts of it, was because of sepsis: wounds were infected with bacteria – it was wartime and the trenches that characterised warfare at the time were filthy, and the whole body became inflamed in an immune response to the infection, and people died; a lot of people died. Fleming, a Scottish pharmacologist, was researching a member of the *Staphylococcus* genera, bacteria that can cause a variety of diseases in humans. He left a Petri dish of the bacteria culture open by accident and went on holiday. When he returned, mould had also started to grow in the dish; there was a bacteria-free area around the blue-green mould, which he subsequently discovered to be *Penicillium notatum*, indicating that it inhibited the growth of the bacterium. While Fleming coined the term "penicillin" for the mould-liquid that he distilled, British pathologist Cecil George Paine was the first person in 1930 to use penicillin to treat medical conditions, which spawned a multibillion-dollar pharmaceutical industry.

This should in no way be considered a recommendation to have a dirty laboratory, but Fleming's accidental discovery changed the course of history, and provided a remedy to what were previously potentially fatal diseases, such as syphilis and tuberculosis, and really horrible ways to die.

A number of Nobel Prizes have been awarded to people who made accidental discoveries. In fact, the first Nobel Prize in Physics went to someone for just such an unlooked-for breakthrough. German physicist Wilhelm Röntgen received a Nobel Prize in 1901 for the accidental discovery of Röntgen waves. In English, these are known as X-rays because he initially dubbed this new radiation "X" for "unknown".

He was fiddling with electric currents and passing them through vacuum tubes – a glass tube from which the air has been sucked out. This was by no means innovative. Physicists had been doing this for a number of years to investigate cathode rays, which were later called electron beams. In November 1895, Röntgen was investigating this phenomenon, with a piece of cardboard covering part of the glass tube to stop the light from escaping. But although there was no visible light, invisible rays were causing a nearby painted cardboard screen to glow. This meant that there was radiation coming from the tube, but it was unknown, an "X" ray. He subsequently used this new discovery to take a picture of his wife's hand – the first X-ray image.

Röntgen later died of cancer, but it isn't believed that radiation caused the cancer. He was one of the pioneers of protective gear, which shielded the wearer from damaging radiation. Unfortunately, his shield was made of lead, which is poisonous and carcinogenic to humans; so it's likely that it was his protective shield that caused his carcinoma.

What these examples seek to illustrate is that the greatest discovery to come out of the SKA project is probably something we haven't imagined. Dr Mike Gaylard of the Hartebeesthoek Radio Astronomy Observatory says that the reason he loves astronomy and why he is still passionate about it even after decades in the game is because there are always new things to discover and so many things we don't know. When the SKA is finally built, there will be balloons and streamers and ribbon-cutting, but the real excitement is what happens in 20 years' time, when we look back and realise just how much we didn't know.

☽

There are areas of science that are opaque to us simply because we do not have the ability to observe them – and this drives scientists and engineers to create new technology to investigate what they previously could not see. It may also drive a number of them mad.

Astronomy provides a particularly frustrating field for the curious, because there is so much uncertainty. A great deal of what we think about the universe, space and galaxies is theorised. As Patricia Whitelock of the SAAO says: "At the moment, we can't go out to a star, take a bit and put it in a test tube and analyse it. We can't give a star a kick and see how it rings. We can't roll it down an incline plane. We can't do any of the things that we, as physicists or chemists, would like to do to find out more about it. All we can do is take the information the universe sends us and analyse that to the best of our abilities." There is no doubt that the machinery we use to receive these universal signals has improved significantly, but as our understanding has grown so has the realisation that there is so much that we don't know, and that the techniques we're using at the moment can only take us so far.

The SKA will be built to investigate five priority areas: the search for habitable planets; the origin and evolution of cosmic magnetism; evidence supporting Einstein's theories of relativity; galaxy evolution, cosmology and dark energy; and, finally, probing the Dark Ages, the time between the Big Bang and the first stars and galaxies. These are rather unwieldy and daunting topics, mainly because so little is known about them. But that is the point of mega-science projects: Go big, or don't bother.

MISSION: Planet hunting
The first, and perhaps least esoteric, plan for the SKA is to search for habitable planets, or "cradles of life". The corollary of this is if these planets are habitable, there may be other life on them. New information – such as data from the Kepler mission, which is tasked with finding habitable planets, and a paper produced by an international consortium, including a South African astronomer – shows that stars in the Milky Way have planets. Importantly, their findings that planet presence is the rule, rather than the exception, show that there are innumerable planets in the universe, which substantially increases the chance of there being small, rocky planets like the one we call home.

However, radio astronomy is an abstract science and, for a number of reasons, it isn't possible to observe these relatively tiny planets directly. The soft radio emissions from these planets would be drowned out by the full symphony orchestra being emitted by its star. Rather like background music that you've tuned out, the Earth emits low radio frequencies, and the source of most of these signals is lightning. With a very low-frequency antenna, you can hear a thunderstorm thousands of kilometres away. While we can hear these signals at that distance, it isn't possible to pick them up if they're light years away.

So the SKA plans to take a roundabout route: to discover where Earth-like planets could be, it will investigate how these planets are formed. Step One is to find young stars like ours. Our Sun is a yellow dwarf, which is a misnomer. To us, our Sun appears yellow because it is filtered through the atmosphere. So stars like ours were called yellow dwarfs, whereas the light they give off is actually white, tending to slightly yellow. The scientific, and less romantic, name is a G-type main-sequence star, which produces large quantities of life-enabling heat through a nuclear fusion reaction in its core, which converts hydrogen to helium. So, first find stars like ours.

By using computer models, the Sun's age has been estimated at 4.57 billion years. Before that, there was just a massive swirling of dust and super-hot gas, and our Sun was born out of it. The rest of the dust – a cosmic smidgeon – coalesced into planets and moons and asteroids orbiting the Sun.

Step Two in the mission to find Earth-like planets is to investigate the dust around yellow dwarfs. This will help us to understand how planets like ours are formed, and from that where we can find them. Yellow dwarfs are estimated to comprise 7–10% of stars in the Milky Way, which puts hundreds of them within 500 light years of Earth, and possibly thousands further afield.

Once again, astronomers have to adopt a circuitous route: star and planet formation takes billions of years. It isn't possible to witness and document the birth of a planet, so instead they need to observe these dust clouds, gathering as much information as possible, to infer how they're formed. A super-duper telescope like the SKA will make that possible, as it will be able to conduct unprecedented thermal imaging of the possibly "habitable zone" in these dust clouds.

MISSION: Cosmic magnetism

Cosmic magnetism is another area in which the SKA is expected to broaden presently limited understanding. Essentially, this is the same effect that you learned about at school with a magnet – how it influences other magnets and the magic of iron filings jumping to attention around it – but take that idea and inflate it to universal proportions. Because of its molten metal core, which is constantly rotating, the Earth's magnetic field runs through its centre, extends out into the atmosphere before returning into the South Pole and then travels through the Earth from south to north. However, the magnetic field changes over time because the magnetised iron alloys in Earth's core are fluid; luckily, the movement is so slow that it doesn't disrupt the compasses we use for navigation. Because the Earth is magnetised, it means that the planet has a magnetosphere, which is formed when streams of charged particles that shoot off the Sun interact with our magnetic field.

"We used to think the Sun was something static that just gave off rays," says Kobus Olckers, space weather officer at the regional space weather warning station for Africa, which is run by the South African National Space Agency. But it is actually "one big nuclear explosion that happens all the time. Like a million fusion bombs going off constantly."

The Sun's gravity is so strong that hydrogen atoms push against each other and form helium, releasing large quantities of energy. This gas expands and creates solar wind, firing off atoms, electrons, photons and radiation of every frequency. The sun also shoots off large lumps of plasma – a super-heated gas formed in a magnetic field. The lumps are very big magnets, often four times the size of Mount Everest. The plasma and solar winds can deform the ionosphere – the charged upper layer of Earth's atmosphere – and the planet's magnetic field. If Earth didn't have a magnetic field, we'd be in serious trouble. Actually, if it didn't have one, we wouldn't be here at all, because it is the magnetic field that shields us from most of the Sun's dangerous radiation.

Not all planets have intrinsic magnets – although Mercury, Jupiter, Saturn, Uranus and Neptune also do – but, at the same time, magnetic fields are present throughout the universe, although they are not uniform. As is the case with the Sun, it is assumed that these magnetic fields are also caused by large swathes of charged cosmic gas, and because they are pervasive they influence the movement of stars and how celestial bodies form and move.

The catch? You can't see them no matter how big your telescope and, unfortunately, we can't scatter iron filings throughout space to show us ordered rows of magnetic field lines. So this is where radio astronomers, armed with an instrument such as the SKA, can make headway where others have failed.

Magnetism can be indirectly detected using a number of cunning techniques. The easiest way is to look for synchrotron emissions, which occur when charged particles travel through a magnetic field. These emissions have a distinct signature and are observable using radio telescopes, although the problem is that not all magnetised objects are energetic enough to produce radiation strong enough to be received on Earth.

Another, even more convoluted, way of detecting magnetism is by observing how a polarised emission from a background object changes when it passes a closer object with a magnetic field, an effect known as Faraday Rotation; in other words, you can detect that there is something magnetised between you and a distant bright object by how the signal changes. Yet, there is a problem with this: the universe is large, infinitely large. Yes, there are many things in it, but the space between them is vast, so there might not be a bright enough background option to employ this technique.

This is an area in which the SKA will show its superiority to every other radio telescope that came before. To use a visual analogy, whereas another telescope will look out into the expanse of space and detect 10 stars surrounded by inky blackness, the SKA will see a dark tablecloth on which someone has knocked over the salt shaker, scattering white grains everywhere. This means that Faraday Rotation will be much easier to pick up with a telescope as sensitive as the SKA.

Then there is the more difficult question: where does magnetism come from? Although we know it to exist and permeate our world and galaxy, we don't know how it originates. Were the magnetic fields there first, ordering how galaxies evolved? Or did they come later, a shifting web of magnetism – and if so, how do they form? And, however they evolved, why do they continue to exist? These are only the questions we have now – who knows what questions the SKA's discoveries will bring in the future.

MISSION: Prove Einstein wrong (or right)

There are a surprising number of question marks over fundamental aspects of the universe. We have developed theories to understand why certain things happen, and the more experiments a theory survives, the more consolidated its position in the annals of science. But an instrument like the SKA will allow us to construct experiments that had previously been impossible in order to test assumptions. A good example of this is gravity.

Albert Einstein postulated a theory of gravity in his General Theory of Relativity in 1916, through which scientists are still trying to poke holes. The more sensitive the equipment, the more ways they can devise to try and disprove this theory. On one hand, this is because scientists are very inquisitive creatures and, on the other, it might be an ancestral hangover from centuries of relying on Sir Isaac Newton's model: For 300 years, gravity was understood in terms of Newton's law of gravitation. In his book *Principia Mathematica*, published in 1686, he described this law in which every object in the universe attracted every other object, and that the strength of this force was linked to the masses of the objects. There is the old wives' tale of his sitting under an apple tree, and this apple knocking the idea into him, but that is assumed to be an embellishment, although Newton maintained that there was an apple involved – even if it didn't hit him on the head. The idea was that there was an attractive force between the apple and the Earth, and that this force caused the apple to accelerate towards the ground.

And this made sense for many years because, on a human scale, this was what we experience. While simultaneously inserting a fly into the ointment of scientific thought, Einstein's theory explained what had previously been cause for confusion – Mercury's orbit. Mercury has the most eccentric orbit of all the planets, and Newton's law of gravitation didn't quite fit with this eccentricity; Einstein's did. In General Relativity, gravity is a by-product of a larger universal system called spacetime. Up until Einstein published his theory, and for a short while after, time was seen as something completely independent of space, and that time continued to periodically tick over, wherever you happened to be – whether at home on your couch, or on Alpha Centauri. Einstein postulated – because remember, all this is theoretical and scientists are still coming up with ways to prove him wrong – that space and time were inextricably linked. Space is usually described as being three-dimensional, and time was thought to be independent of that relationship. But Einstein changed all that by saying that time was the fourth dimension in the spacetime continuum and just as necessary as any of the space dimensions.

Everything in the universe has a position in space, but also in time, and both are dependent on each other. While you may not observe that going about your daily life – even if you are convinced that time behaves differently in certain situations, such as when you're on deadline and it speeds past your ears, or when you're in the middle of an awkward conversation and seconds become hours – it becomes obvious when you're travelling near the speed of light.

The long and the short of it is that space is curved. Think of two people at opposite sides of the equator who start walking towards the North Pole – they will eventually meet, not because there is a force drawing them together, but because the planet is curved. The same idea can be extrapolated into the rest of the universe, explaining the motion of planets and stars. Einstein said that objects – such as planets and stars – distorted this continuum, which means that other bodies behaved differently when they were caught up in that object's distortion of spacetime. So gravity didn't exist as a force that caused planet A to rotate about planet B; rather, planet A took that route because it was the most expedient way to navigate the spacetime distortion around planet B.

The problem with theories is that they can be like the scientists who come up with them: contradictory. Einstein's General Relativity seems to work and you can apply it to different situations, but the same can be said of quantum theory, which we use to explain the behaviour of things on a subatomic scale – and the two don't always agree on what will happen. So scientists need to investigate the matter, and the SKA will be a handy weapon in their arsenal.

Gravitational radiation was first confirmed 30 years ago by observing radio pulsars – and earned Joseph Hooton Taylor Jnr and Russell Hulse well-deserved Nobel Prizes in 1993 – so the best place to start is with pulsars. A pulsar develops after a massive star has gone supernova, a cosmically impressive and gigantic explosion, and has collapsed into a neutron star that

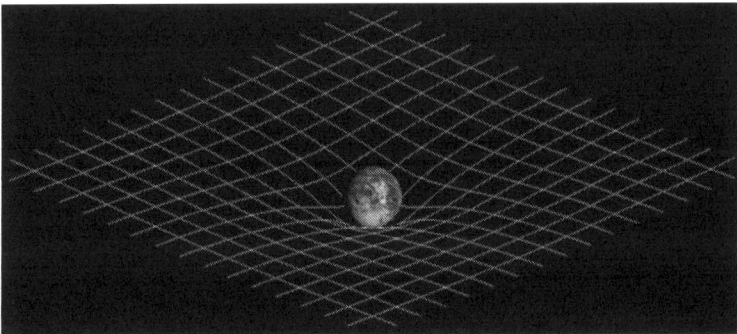

Spacetime continuum

continues to rotate. Neutron stars are incredibly dense and highly magnetised, while being very small – an object about 20 kilometres in diameter with a mass about 1.5 times that of our Sun. An interesting property of pulsars, and what makes them the darling of the radio astronomy world, is that they emit a strong beam of electromagnetic radiation from their axis of rotation. So if Earth was a pulsar, it would be smaller, made up of neutrons, all of us would be dead, and it would shoot beams of radiation out of the poles.

Astronomers can only see this emission when the beam is facing Earth, and since the pulsar is rotating, it can only be seen periodically – like a lighthouse. Some pulsars are so accurate in their rotation that they rival atomic clocks – the most accurate time-keeping device on Earth – for precision. Now, because the signals from pulsars are very regular, it means that you notice when they deviate from what you expect – this was how gravitation radiation was confirmed. Because the SKA is so sensitive and will have such an unprecedented view of the sky, it will be able to scour the skies for pulsars and detect minute deviations in their behaviour. First prize will be if astronomers discover pulsars orbiting around supermassive black holes, which are regions where matter is so dense and the gravitation field so strong that not even light can escape from it. By monitoring pulsars orbiting such a region, it will be possible to investigate electromagnetic emissions in strong-field gravity conditions. From this, we will either confirm Einstein's theory or devise something new to replace it.

It must be noted that black holes are another one of the things postulated in Einstein's theory of relativity that no one has actually proven. Since even light can't escape a black hole, you can't observe it directly – you can infer that it is there by how other objects behave around it. There have been a fair number of candidates identified, though. So the black hole theory is also up for inspection on the SKA's to-do list.

MISSION: Life, the universe and everything else
The SKA will perhaps give us answers to the fundamental questions that torment scientists and terrify parents with young children: How was the universe made? What is the Big Bang? Why is there something instead of nothing? The answer of both parties is, "Well, we're not quite sure."

Scientists have theories to explain some of them, but at this stage that is what they are: theories, a hypothetical model that seems to plug some of the holes, and explains what scientists observe.

There is a misconception that science is infallible – that it's a series of axioms, carved into stone – whereas it is much more of a work in progress. A theory is only as strong as the observation that disproves it. It works like this: a scientist (for simplicity it can be one, although there are often teams of big brains working on a single problem) observes that under a certain set of circumstances something happens. In order to explain why that thing happens, he or she devises a model and a theory to explain it. If another scientist devises an experiment that disproves the model, well then that theory is relegated to a dark corner, or the mountains of discarded theories that litter the history of science. These mountains are substantially bigger than the pile of ones that actually work.

And the Big Bang, the prevailing model for the universe's development, is just that – a theory, one that every day is subjected to the merciless prods of scientists. The SKA will be the most effective "prodding" tool we've had to date, although "prodding" might be a euphemism – it's more like removing all the organs from your body to see how it works.

An estimated 14 billion years ago, the universe – the galaxies we can see, and the ones we can't, all that matter swirling – was concentrated into a very small and very hot space, known as the singularity. For some reason, this singularity exploded in a cataclysmic event and started accelerating outwards. Think of it like this: you have a small bag and a lot clothes, and somehow – in an apparent contradiction of the laws of physics – you manage to cram all the clothes in the inadequate bag. When you open the bag, the clothes leap for freedom and jump across the room. It's sort of like that, but on an unimaginably large scale, far beyond the abilities of a humble bag.

After this grand explosion, the universe began to accelerate away from the scene of the crime. It was so hot that it wasn't even matter, it was pure energy speeding through chasms of space. It cooled as it moved further away, and began to develop into protons, electrons and the like. The next step in this microscopic evolutionary process was for these subatomic particles to organise themselves into atoms, which is thought to have taken thousands of years. And the simplest atom? Hydrogen: one proton, one electron. Atomically, you don't get much simpler than that.

The key to understanding galaxy formation is hydrogen, the first element on the Periodic Table and the most common chemical substance in the universe. When hydrogen is hot and has a lot more energy – like it would have been in the early days of the universe – it behaves slightly differently to when it is cool. This energetic hydrogen is called neutral hydrogen,[1]

---

1 Neutral hydrogen (which is made up of one hydrogen atom) is not the same as molecular hydrogen gas (which is two hydrogen atoms H_2).

while your less energetic and more lethargic version is known as ground-state hydrogen.[2] When it cools, becomes less excited and changes from neutral hydrogen into ground-state hydrogen, it emits radiation – a characteristic wave with a wavelength of 21 centimetres and a frequency of 1420 MHz, otherwise known as the hydrogen line.

The complicating factor – as though this sort of science needed anything to make it more complicated – is that the universe is expanding, distorting the signal – well, that is what is theorised under the Big Bang theory. This is called the red shift, and is what astronomers search for to detect this radiation. The sound of an ambulance siren is distorted and stretched as it moves away from you. The same thing happens with the hydrogen signals, except over light years. This stretched signal phenomenon is called a red shift because red light has the longest wavelength – this is astronomer shorthand for the elongation of an electromagnetic wave, rather than the signal being in the visible spectrum.

If the universe was contracting and matter in the universe was accelerating back towards the origin of the Big Bang, it would be called a "blue shift", because the signals would become scrunched – like an accordion – and the wavelength would be shorter, and blue light has a shorter visible wavelength. To return to the ambulance siren, if it was travelling towards you, the sound would become more high pitched as it got closer.

The red shift – which has been observed – is one of the strongest pieces of evidence in the Big Bang theory's favour. The other is the cosmic microwave background radiation: throughout the universe, there is an almost uniform thermal

---

2 For those who are interested in more detail, the proton and the electron in hydrogen are spinning within the atom. When the atom is more energised, the "spin" of the proton and the electron are in the same direction; in other words, they are aligned. After the characteristic radio emission, the proton and the electron "spin" are anti-aligned and spin in the opposite direction to each other.

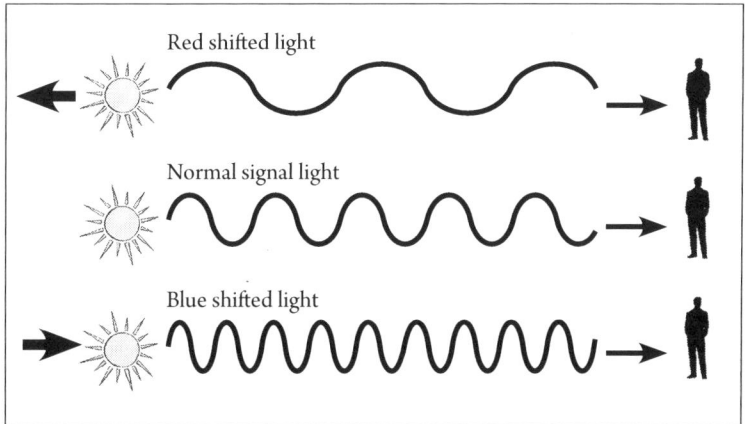

Red shift vs normal signal vs blue shift

radiation – about 2.7 Kelvin.[3] This strange observation can be explained using the Big Bang theory. Back in the early days, once the cosmic primordial soup had congealed into something more tangible – when energy coalesced into subatomic particles, which merged into atoms in what is called the Epoch of Recombination – the universe was opaque. This was known as the Dark Ages – well, the cosmic Dark Ages rather than the period following the fall of the Roman Empire when humans carelessly lost writing, mathematics and a host of other useful intellectual attributes. Following the Big Bang – if we assume that theory for the moment – everything was very, *very* dense, so although there were photons (light particles) swirling around in this soup, the distances between particles were so small that the photons kept on knocking into other particles. It is kind of like being at a packed party and you are trying to get across the room to chat to the pretty girl on the other side, but you keep on bumping into people you know – except in this instance you know everyone.

---

3 A Kelvin degree is the standard unit of temperature, and has the same magnitude as degrees Celsius (except for being +273 more). So, for cosmic microwave background radiation, it is the same as the entire universe being 2.7º C warmer than you expected.

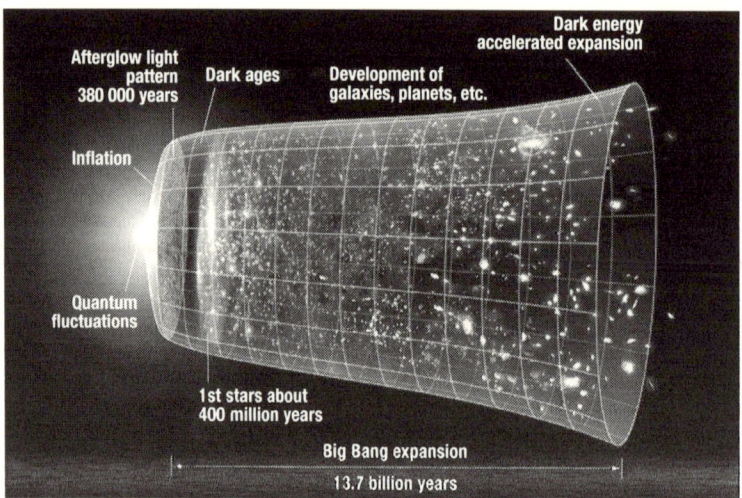

The history of the universe, as hypothesised by the Big Bang theory

Only once hydrogen, and then larger atoms, started to form could these photons escape and travel longer distances, which made the universe transparent. This light has been travelling for 13 billion years, and so when it reaches Earth, it is no longer the energetic light we know, such as the visible light that comes off the Sun, but has been stretched into microwaves, which is why it is called cosmic microwave background radiation.

Once again, this is all theory, although it is the best one we have at the moment.

The question is then: how did the galaxies and stars form? This is called the Epoch of Reionisation, when the first stars and galaxies were formed. The SKA's main task will be to "see" back in time and space to this Epoch of Reionisation, mainly because we don't really know how galaxies are formed.

A galaxy is a huge system of stars, planets, dust and gas. A small dwarf galaxy has about 10 million stars, while a giant galaxy can have up to a hundred trillion stars. Our Milky Way has about 200–400 billion stars; our Sun is just one of them. For some context, and a South African example, take a R2 coin.[4] If you were to stack 10 million (seven zeros) of these coins on top of each other, your coin tower would be 20 000 kilometres high – and if you were to lay it on the ground, it would run from pole to pole. If you were to stack a hundred trillion (14 zeros) coins, you'd be very rich. You would also be able to wrap the coin chain around the Earth and Venus – and the great distance between them, when it is at its maximum (the Earth–Venus distance varies with their orbits) of 261 million kilometres – 380 times. Yes, it is really that big.

The main element within these galaxies, whether big or small, is hydrogen, and in order to understand how those first galaxies formed in the Epoch of Reionisation, we need to know how they form now. But it isn't possible to study a galaxy from birth to death, our insignificant Sun would have long since kicked the bucket – estimated at about five billion years from now – and swollen into a red giant, when all the hydrogen in the Sun's core had fused into helium. The idea behind the hydrogen fascination is that astronomers believe that these atoms are the building blocks of stars. In the same way that our Sun squeezes hydrogen atoms together to create helium, all the elements in our universe – up until lead, which has 82 protons (think of 82 hydrogen atoms squeezed together into one) – were created through the nuclear fusion that takes place in stars. Atoms with higher atomic numbers are created through supernovas, giant stellar explosions.

So, first, identify where the hydrogen is and what it is doing. To figure out how galaxies are born and die, and by extension proving or disproving the Big Bang theory, the SKA will start by looking for hydrogen. Because hydrogen atoms emit a constant

---

4 Dear confused international reader, a two-rand coin is about 1.4 centimetres in diameter and 0.2 centimetres in width, with a silvery shine and a picture of a Greater Kudu.

and identifiable frequency, it is possible to study these galaxies, and it is surprising how much you can learn about them based on these hydrogen-atom signals. First, you can figure out whether a galaxy is moving away or towards Earth, through slight alterations in the usually constant frequency of hydrogen, through what is known as the Doppler Effect.[5] You can also tell how fast the galaxy is rotating. Radio astronomers can detect – and with the SKA it won't be so much "detecting" as "difficult to miss" – this alteration. If you know how fast a galaxy is spinning, you can deduce how big it is. Now, an object's rotation is linked to how much mass there is and how it is distributed. In the classical Newtonian framework, mass attracts other mass to it and so a galaxy sticks together; in Einstein's view of gravitation, the matter, caught in gravitational radiation, remains in relatively close proximity.

The salient point is that stars cluster together to form galaxies, and their rotation is linked to their mass. But sometimes the figures don't add up: the hydrogen may be spinning very quickly, leading an astronomer to deduce a certain mass, but this mass seems much larger than it should be when the constituent stars are taken into account. Optical observations tell us how much light there is, and therefore how much matter, but the rotation of the galaxies indicates that there is actually more matter. And this is where dark energy and dark matter come in. Well, either that or Einstein was wrong (see also "Mission: Prove Einstein wrong (or right)" on page 160).

In terms of the SKA's to-do list, searching for hydrogen is perhaps the low-hanging fruit in their scientific quest. Dark energy and dark matter? Well, that really is the search for the holy grail, because it would explain why the universe appears to be accelerating, why galaxies hold together even though it doesn't seem as though they should. It must also be said that "the SKA is searching for dark energy" sounds much more exciting than "the

---

5 The Doppler Effect causes the frequency of a wave to alter as it moves.

SKA is searching for the most common chemical element in the universe". This may seem more fascinating because dark matter has a catchy name. It would probably be less intriguing if it was called "something that permeates most of the universe, that we're sure is there, but we just can't put our finger on", although both descriptions would be true.

The idea that the universe is expanding is not new. Astronomers first started to get an inkling of it in the early 1900s, which in 1927 led the Belgian physicist and Catholic priest Monsignor Georges Lemaître to propose the Big Bang theory. But it was only in 1998 that three physicists – Saul Perlmutter, Brian Schmidt and Adam Riess – discovered that the universe wasn't just expanding – it was accelerating.[6]

But no one knew why. This is where dark energy and dark matter come into it. According to the present cosmological model, 74% of the universe is dark energy, 22% is dark matter and the remaining 4% is everything else, of which 3.6% is intergalactic gas. So the stuff we can see is only 0.4%, which – more than anything else – gives you a sense of just how big the universe is.

Dark energy is thought to be distributed uniformly throughout the universe and explains why the universe is expanding. It exerts pressure on matter and interstellar gases, forcing them to accelerate – a sort of anti-gravity. Dark matter does the opposite: galaxies can form and stay together – even though there isn't enough visible matter to justify this cohesion – because dark matter is attractive.

Bear in mind, all this is hypothesised. There is some data to back it up, but no one is certain. Dark energy and dark matter can't be seen directly, but their presence can be detected by how radio waves behave around them. Using our present techniques, it's a bit like playing Blind Man's Bluff, a game where someone is blindfolded and has to wander around a room hoping to bump into someone else.

---

6 They managed this groundbreaking feat through studying distant supernovas, otherwise known as enormous exploding stars.

There are other theories about the behaviour of the universe, although fewer since the discovery of expansion. Some people believe it is in a steady state, neither expanding nor contracting, and definitely not accelerating. Others believe we live in a multiverse and our universe is only one of an infinite number of other universes, like one card in an infinite pack.

And how do you deal with so much scientific uncertainty? You build the most impressive instrument you can – constrained only by ability and imagination, not to mention the money – to answer the questions that no one else could.

> **The girl who made the stars**
> The Sun had set and the Moon had begun its furtive flight across the sky, and a girl was hungry. She lay in her mother's hut, feeling hunger's teeth gnaw at her stomach and becoming more irritable as the minutes stretched into hours.
>
> She was waiting for the return of her mother, who had gone out to collect roots for her to eat. But all she wanted was a piece of meat, kudu or springbok, roasted in the fire. The thought of the slightly crispy flesh made her mouth water and her stomach shout at her angrily. But because she was now a young woman, only her father could kill game for her, and her father was old and was no longer a good hunter. She couldn't eat young men's meat, because then her saliva would seep into the meat, and then the arrowhead of the young man's bow would grow cold. This frigid cold would seep into the bow and all the way up the man's arm. And then no one would be able to eat meat, because the young man would not be able to shoot animals with a cold and useless arm.
>
> If only it were brighter, the girl thought. Then my mother would be able to come home faster and I could eat, and perhaps my father would shoot something and then I would be able to feast. Her ears

strained to hear the slightest rustle, and she heard the women lumbering through the veld even though they were still far away. When she saw the meagre forage, her heart was heavy and her temper was hot.

She watched as her mother put the roots into the fire, pushing them into dark crevices between the glowing coals. "Don't stare at the fire like that, you will scare it into coldness and then none of us will eat, you ungrateful girl," her mother chastised. "The night was dark and we collected all the roots that we could. When you go out to seek food, you will understand."

In a fit of rage and hunger, the girl reached into the fire, not caring if she burnt her hands, picked up the roots and wood ashes, and threw them into the sky.

"Wood ashes and roots, go up into the sky and make the world light, so that the men hunting and the women looking for food will be able to see the animals and the roots, and come home quickly," the girl said. "Turn with the sky so that there will always be enough light to see, and remember that you were once embers and burn away the darkness."

The pieces of root heard the girl's words, and began to glow red and white, while the ashes fell into a path across the sky. Some people call the path the Milky Way and others the Stars' Road. While the stars travel to fetch the Sun each night, and remind him to rise in the East, the ashes sail across the heavens and guide the foragers' footsteps home. They cast their light even when the Moon is not there, because the embers and roots remember the hungry girl and they still do what she asked.

12
Looking to the future

All this talk of black holes and the beginning of the universe usually gives rise to the same response: "Yes, that's all very interesting, but it's not going to change my life. Why should I give a damn?" You should care because all of the nifty equipment and technology developed to divine these answers will eventually benefit you on a practical, everyday level.

For example, Nasa's Apollo missions gave rise to technology that you use daily without knowing who was responsible. Today's running shoes are derived from the first boots that walked on the moon. For the Apollo missions, astronauts were clad in specially designed space suits, which included – you guessed it – moon boots. These shoes literally put a spring in your step, while providing ventilation for your feet. Sound similar to running shoe adverts? It should, because that was what this technology finally produced. In the mid-1980s, shoe company KangaROOS USA, with the help of Nasa, patented a three-dimensional polyurethane foam fabric for the inner sole of the shoe. It distributes the force of the foot on the ground and absorbs part of the impact. This is why athletics shoe companies can offer you shoes that minimise the impact on your joints.

Similarly, those of you who wear glasses have Nasa to thank for the fact that you don't have to replace them every couple of months because of scratches. If you think scratched spectacles are inconvenient, imagine being an astronaut in space, surrounded by dust and particles that damage your visor and make it almost impossible to see. Unlike glass, plastic is cheaper, better at absorbing ultraviolet radiation and is less likely to shatter, but uncoated plastic is also very prone to scuffing. This is why Nasa developed a special coating technology that made this plastic scratch resistant.

The best thing about these sorts of developments is that they are seldom planned for. Projects such as Nasa's Apollo mission and the Square Kilometre Array push the boundaries of what people think is possible – they have to, because they are doing things that haven't been done before.

Sandile Malinga, the head of the South African National Space Agency, once spoke to me about how South Africa could justify a space programme, given its desperation for basic services – a similar argument to the one used against the SKA. "It's central to the business of the country – space is a natural innovation driver. It forces you to break away from the normal … I mean, the International Space Station, to build that kind of gigantic thing up there, you have to build things differently," he waves a hand in the direction of a corner of his office. "It's not like you're constructing it here in my office. That's easy. So here is this natural innovation we can use as a tool for training and skills in different areas."

The same thing can be said of the SKA, which will push the boundaries of what we can imagine. Right now, we do not have the technological capabilities to realise the kinds of things that we want the SKA to do.

Let's start with the sheer quantity of raw data that the SKA will produce. If you were to play on an iPod the data collected in one day, it would be two million years of listening time. Sticking with the iPod analogy, the raw data generated in a day will be able to fill 15 million 64GB iPods.

To do that kind of science, you don't just have a computer. You have a Super Computer, a computer that at times you worry is smarter than you; one that brings you your coffee in the morning, sourced overnight from a plant that is only found in a square-kilometre radius in the deepest Amazon, whose beans you can only pick at full moon. That kind of computer. This central computer will be the equivalent of one hundred million PCs – that's *eight* zeros.

The data collected from the dishes and the aperture arrays will be transported, via fibre-optic cables, to this Super Computer. In fact, the fibre-optic cables connecting all the different parts of this huge interferometer, if taken as one strand of cable, will be able to wrap itself around the Earth twice. Then the data will be correlated, calibrated and analysed – in other words, it will be added together and turned into a signal that radio astronomers recognise. This is all written in the future tense because at the moment this technology doesn't exist. There are systems that can do these sorts of computations at a more basic level, but not to this degree. Even if you just take the 3000 dishes, which are only one part of the SKA, each dish will transmit approximately 160 GB of data per second. This is the same as two and a half 64GB iPods per second. Taken all together, the dishes will produce 10 times the amount of traffic on the Internet. Add in the aperture arrays, and this is about 100 times the global Internet traffic.

Naturally, industry wants in on the action, both for the

benefits that this sort of research and development can lend their existing business – whether they are technology companies, signal broadcasters, mobile service providers, signal and data processors, receiver developers, infrastructure specialists – but also to get a foot in the door of future industries. "The new industries that will become major industries in the next 10, 20, 30 years … [will] increasingly focus on large volumes of data, the speed at which you can transmit them and the problems of computing them," says Bernie Fanaroff.

The modern world, with its smart phones, laptops, social networks, multiple servers and the Internet, needs to be able to cope with large quantities of data: to transport and process all this data as well as turn it into information. The global economy is now dependent on information, which is different from data. Data on its own isn't useful. It has no context – it might just as well be a series of ones and zeroes on a page. Data begins to have value when it is turned into information, which is data that has been condensed and organised into trends and patterns. "There is a flood of information and data, and converting the data into information, and the information into understanding is a major challenge," Dr Faranoff says. "Those are going to be new industries that didn't exist in the past because it just wasn't something that was an issue."

On the one hand, our Information Age necessitates that we are able to store and process large quantities of data, and that we can turn this data into information quickly. Sound familiar? That is exactly what the SKA will have to do.

While this makes it a tantalising opportunity for businesses, it means that in the long run it will ultimately affect your life. If Intel, IBM, Nokia Siemens – or any of the other industry partners for that matter – develop a way to offer consumers faster computing or improved systems, these technological innovations are likely to literally change your life. Think of Nasa and its moon boots.

Since the SKA is such a complex process, there are many

industry partners working on various aspects of it[1] – globally as well as in the two partner countries.[2] In one particular aspect of SKA technology, South Africa has turned into something of an expert – computer processing. As part of its human capital development project for the MeerKAT and to bolster its SKA bid, SKA SA sent South African engineers to international organisations and institutions to learn more about engineering for radio astronomy and to give them a taste of the big leagues.

The University of California Berkeley has a project called Casper – the Collaboration for Astronomy Signal Processing and Electronics Research – which South African engineers slotted into. The Casper programme aims to create an open-source hardware and software system – the physical computer parts and the codes written to make them operate – that could be used on any radio astronomy instrument to streamline the flow of data and the creation of information. The data from the telescopes is fed into a correlator, which transforms the data into a signal presentation. This sort of technology will be used for South Africa's MeerKAT and may also be used for the SKA.[3]

As Francois Kapp, subsystem manager at MeerKAT, says: "The MeerKAT processes about a terabit [of data coming from the array telescopes into the correlator] per second. A megabit

---

1 The technological innovations and SKA spin-offs deserve a book of their own, possibly more than one. Rather than pretend to do them justice, this book will cherry-pick and look at two examples.
2 The great thing about being a journalist is that you get paid to ask what some may consider very rude questions. There have been many press briefings announcing a collaboration between SKA SA and a global industry partner. The first round of questions contains the usual "who is involved, how much money and why are you doing this?", etc. After the ice has been broken, it is time to ask the question everyone is thinking, but no one wants to ask: "So, are you also working with Australia?" Throats get cleared and CEOs look to their PR people for guidance, not wanting to admit that while they are proudly collaborating with South Africa, they are – in most cases – also proudly partnering with Australia.
3 It has still not been decided which technology will be used for the SKA.

per second is an ADSL line, so you're talking about a million ADSL lines."

Enter ROACH, otherwise known as Reconfigurable Open Architecture Computing Hardware. Designed in collaboration with the University of California Berkeley in the US, ROACH is "based on field-programmable gate array (FPGA) technology: a specific kind of chip, a blank canvas with a sea of gates that you can connect any way you like," says Kapp.

"The aim is to build a PC for radio astronomy, with generic processing elements, and then load a processing function on to them, and then hook them together." The result is "an instrument [that] is tailor-made, but consists of generic building parts", explains Kapp. This first-generation ROACH-1 technology is already in use at the KAT-7 and has piqued international interest, with more than 300 science facilities employing the ROACH-1 boards.

And this is only the beginning.

The second-generation board – the ROACH-2 – is much faster and more powerful. But future incarnations of the ROACH boards will have to be even faster and more powerful.

The prototype KAT-7 comprises seven dishes; the MeerKAT project will be an array of 64 dishes; and the final SKA thousands. Kapp says: "For SKA Phase 1, there are 250 dishes planned. This is four times the number of the MeerKAT. [But] because it's a correlator, you need 16 times the processing capacity. For the SKA proper, we'll need 10 times the data rate of Phase 1, which is 100 times the data processing." Luckily, Moore's Law is in our favour, Kapp says.

Moore's Law, as explained in Chapter 11: SKA Science, is a long-term trend in the history of computer hardware, which projects that data-processing capacity will double periodically while the price remains the same. In fact, a great deal of the ambitious SKA technology is based on Moore's Law because at the moment the technology doesn't exist to accomplish the SKA's rather daunting computational tasks. If Moore's Law holds and SKA scientists and engineers incorporate these advances into their designs, the technology will exist in the next decade. "So we can project where it's going to go, and know where it's going in the future," says Kapp.

Naturally, with this kind of processing capacity even in its nascent ROACH-1 and -2 guise, the private sector is interested. "There's a great deal of interest and excitement from industry; we're getting inquiries about using the tools in non-radio astronomy activities," Kapp says, without giving names due to confidentiality agreements. And this is not surprising, considering the fields in which this technology would be useful.

Kapp cites genome sequencing. "The FPGA is uniquely suited to that as it can cater for the four [genome] letters and pack it into that space. There will be spin-offs that will affect people directly. Genome sequencing that takes 10 days can be done in an hour or two. It becomes something you can do on a wide base. There are many large medical applications," he says.

Not to mention telecommunications. Faster data processing means increases in the speed at which we communicate, be it through wireless or radar. He chuckles. "But we're going to try to focus on the telescope, because it keeps us busy enough."

The other side of this is what to do with the information once you have it. That is where collaboration with global technology companies comes in, such as between the SKA and IBM technologies. "We want less human input and more artificial intelligence software to assist with the calibration of data we're producing," says Jasper Horrell, who is part of SKA SA's MeerKAT science processing unit. Calibration is "understanding and correcting for the effects of the instrument itself, which gets in the way of the science". Doing this manually is a nightmare, even on a "normal" and not gigantic radio telescope. You would need many people, working in ceaseless rotation, to be able to calibrate all the data that the MeerKAT will produce, ignoring the SKA. There is also the fact that it is grunt work, and highly unenjoyable. The author can personally attest to this – days of staring at Excel spreadsheets, trying to divine meaning from series of columned numbers and then going back and doing it all again when you realise that you didn't correct for the vagaries of the radio dish itself. Saying it is unenjoyable is so euphemistic that it is basically lying.

The partnership between IBM and SKA SA is in its early stages, and aims to combine machine-learning techniques under development at IBM Research and radio astronomy analysis software, effectively making a platform with self-correcting capabilities. "The goal of the proposed project is to teach a computer to make perfect images on its own," IBM researcher Alain Biem says. "A software platform like this may assist in enabling large survey instruments like MeerKAT to process the trillions of bits of data per second they receive and make it available to astronomers around the world."

It starts with radio astronomy. These innovations begin life as ideas on paper. They morph into prototypes; they shift into one-of-a-kind scientific instrumentation, to be copied by scientists and engineers at other academic institutions. At each step, this concept – be it intelligent software, composite material antennae, super-fast processing, Super Computers, cryogenic freezing systems to keep the receivers cool – pushes the boundaries of what had previously been thought possible, and then you look back at the paper trail that has solidified into technological innovation – and imagination and the crazy concepts of science fiction are suddenly all around you, in a world that 20 years ago no one had thought possible.

If this is where we are now, just imagine what we will know when the SKA is finally operational in 2024. We will look back and marvel at just how much we didn't know.